Littorio
VS
Richelieu

Written & Illustrated by Paul Forest

Copyright © 2021 Paul Forest
All rights reserved.
ISBN: 9798589911398

Introduction

In 1929 the Germans laid the keel of the 11.14" (283 mm) *Panzerschiffe* (armoured ship) Deutschland, followed by two more units of the same type in the next three years. The French countered the incurring threat by the 12.99" (330 mm) battlecruisers Dunkerque and Strasbourg, laid down in 1932 and 1934, respectively. The Italian reply to French naval expansion was to begin the construction of two 15" (381 mm) battleships, Littorio and Vittorio Veneto, in 1934. The French responded in kind, laying the keel of the 14.96" (380 mm) battleships Richelieu and Jean Bart in 1935-1936.

Littorio and Richelieu were commissioned during the spring of 1940. Although the imminent Fall of France prevented the envisaged struggle for the control of the Mediterranean between this nation and Italy, the climax of this campaign would have been undoubtedly a gunnery duel between the newly commissioned battleships. We now endeavor to investigate which ship would have been more likely to emerge victorious from this hypothetical engagement.

Contents

Introduction 5

Abbreviations 9

Chapter 1 – Firepower 11
    Littorio's Main Battery 11
    Richelieu's Main Battery 13
    Comparison of Main Batteries 15
    Medium Caliber Guns 24

Chapter 2 – Fire-Control 25
    Littorio's Main Battery Fire-Control 25
    Richelieu's Main Battery Fire-Control 27
    Comparison of Main Battery Fire-Control Systems 29

Chapter 3 – Protection 35
    Littorio's Protection Scheme 35
    Richelieu's Protection Scheme 38
    Comparison of Protection Schemes 41

Conclusion 71

Appendix A – General Characteristics 75

Appendix B – Calculating Penetration Values 77

Appendix C – Decapping Plates 81

Appendix D – Richelieu's Secondary Armament Fire-Control System 85

Appendix E – Jean Bart's Guns and Directors 87

Appendix F – U.S. Radars of Richelieu 93

Appendix G – Projectiles 99

Appendix H – Jean Bart's Protection Scheme 101

Appendix I – Turret Mounts 107

Appendix J – Longitudinal Sections 109

Appendix K – Major Caliber Hits Scored Against French Capital Ships 111

Appendix L – Official Penetration Capacity and Gunnery Trials of Littorio's Main Guns 113

Definitions 115

List of Tables 119

List of Figures 121

List of Pictures 123

Bibliography 125

Abbreviations

AA – Anti-Aircraft
ADM – Admiralty Minutes
AP – Armour Piercing (projectile)
APC – Armour Piercing Capped (projectile)
APCBC – Armour Piercing Capped Ballistic Capped (projectile)
ATC – Armour Technical Committee
Bhd. – Bulkhead
BHN – Brinell Hardness Number
Bu. Ord. – Bureau of Ordnance
C – Cemented (armour) or Coincidence (rangefinder)
cal. – caliber(s)
CT – Conning Tower
deg. – degree(s)
DEM – Détection Électro-Magnétique (radar)
Fwd. – Forward
HE – High Explosive (projectile)
HTS – High Tensile Steel
IV – Initial Velocity
KC – Krupp Cemented (armour)
Mle – Modéle (model)
NC – Non–Cemented (armour)
NiCr – Nickel Chromium (French NC armour)
NPG – Naval Proving Ground (Dahlgren, Virginia)
OD – Omogenee Duttile (Italian NC armour)
OPfK – Obus de Perforation dispositif 'K' (French APC with dye bag)
OPL – Optique de Précision (precision optics) Levallois-Perret
OTO – Odero-Terni-Orlando (Italian company)
R.E. – Relative Effectiveness (explosives)
RF – Rangefinder
RPC – Remote Power Control
S – Stereoscopic (rangefinder) or Special Steel
SAP – Semi-Armour Piercing (projectile)
Sk. – Sketch
T. – Torpedo
UTS – Ultimate Tensile Strength

Chapter 1 – Firepower

<u>Littorio's Main Battery</u>

Littorio carried the nominally most powerful guns of this diameter (15"/381 mm) ever mounted on a battleship. This was due to the fact that the Italian guns were longer (50 cal.) than any other of this diameter carried afloat and had exceedingly high average pressure. The latter of these characteristics had the unwelcomed side effects of rapid bore erosion and great dispersion, however. In order to remedy these shortcomings, pressure was slightly reduced, although it remained well-above foreign standards.

Based on gunnery exercise reports, the average deflection error was 1.70 % of range, range error 1.84 %, at the optimal battle range. Even though these values are inferior to foreign ones – these usually range between 1.0 % and 1.5 % of range in cases of 3-5 gun salvos, assuming firing delays and the exclusion of wild shots – they are acceptable.

The Italians utilized the great muzzle energy of these guns to propel projectiles of slightly above-average weight (0.58 lbs/in^3) at very high initial velocity – 2,854 fps (870 mps), later reduced to 2,789 fps (850 mps) – which resulted in unparalleled ballistic characteristics. Indeed, at an elevation of 36 deg., the range of these guns was unmatched.

The combination of great terminal energy and flat trajectories induced extreme vertical armour penetration capacity. More concretely, Bu. Ord. Sk. 78841 indicates that the Italian gun was superior to all 16" (406 mm) and smaller guns afloat in terms of vertical armour penetration capacity, with the sole exception of the 16" (406 mm)/50 guns of Iowa, which had a marginal edge in this regard.

However, impact angles against horizontal armour were very disadvantageous. As a consequence, heavy deck armour could be pierced only at extremely long ranges.

Other shortcomings of the Italian guns included their unimpressive rate of fire. Based on empirical data, the sustainable rate of fire was circa 1.3 rounds/gun/minute. Also, the AP type projectiles carried surprisingly small quantities of explosives, as only 1.1 % of their total weight was reserved for this purpose. The filling material was TNT.

Picture 1.1

The battleship Roma displays her armament.

Richelieu's Main Battery

The modern 14.96" (380 mm) gun of the French navy was only 45.41 cal. long, but its extraordinarily high average pressure made it a very powerful weapon for its caliber. Indeed, muzzle energy rivaled that of a low-pressure 16" (406 mm)/45 gun, which enabled it to propel relatively heavy (0.58 lbs/in^3) projectiles at remarkably high velocity (2,723 fps/830 mps).

In order to minimize retardation, the French designed very long (5 cal.), boat-tailed shells which, combined with a high initial velocity and sectional density, ensured splendid ballistic characteristics.

As in contemporary Japanese designs, considerable emphasis was placed on enhancing the underwater performance of projectiles. In French designs, this was attained by separating the ballistic cap of shells into two parts, namely a tapering forward section containing a dye bag and a lower section attached to the armour-piercing cap. Upon contact with water, the upper section broke off and the resulting flat contact surface allowed the projectile to continue its trajectory submerged relatively undisturbed.

Amounting to only 9.5 % of their total weight, the AP type projectiles had relatively light armour-piercing caps. Furthermore, while foreign navies preferred blunt projectile body tips and armour-piercing caps to increase bite angle, those of French manufacture were comparatively tapering, especially at the sides, which might have decreased their efficiency at high obliquity, although Bu. Ord. Sk. 78841 credits them with extremely good vertical armour penetration capacity as a result of their great terminal energy and flat trajectories. The shape of projectiles and the flatness of their trajectory, however, had a detrimental effect on horizontal armour penetration capacity.

The AP projectiles incorporated bursting charge containers of considerable volume (2.5 % of total weight). The explosive material was composed of 83.2 % picric acid and 16.8 % dinitronaphthalene. The former is a sensitive but powerful explosive with higher R.E. than TNT. The addition of the latter, a relatively weak but insensitive explosive, was necessary to prevent premature detonations upon contact with the exterior of the target.

The rate of fire turned out to be below expectations. While it was envisaged that the loading cycle would be completed in about 30 seconds, it took circa 45 seconds to reload the guns. This was partially caused by insufficient replenishment system arrangements, which were later improved.

Although barrel life was acceptable, combined with faulty projectile design, high internal pressure caused failures at Dakar. This was remedied by the reduction of pressure and the reinforcement of the base of projectiles.

After the U.S.'s refit of the battleship, initial velocity dropped to 2,625 fps (800 mps) for a new gun and 2,575 fps (785 mps) for an average gun.

The 14.96" (380 mm) projectiles manufactured by the Americans were negligibly shorter (4.95 cal.) and contained slightly less explosive (2.3 % of total weight). Also, the French picric acid/dinitronaphthalene mixture was abandoned in favor of explosive D, which has an R.E. of circa 0.95. The weight of the U.S. projectile was the same as in the French design, but the shell body tip was blunter and it was topped by a much more robust (14 % of total weight) blunt AP cap. Hardness distribution was most likely the same as in contemporary U.S. designs, that is, based on NPG 3-47, a sheath hardened projectile body with a max. hardness of about 550-580 BHN with a cap of the same max. hardness.

Owing to the reduction of initial velocity, ballistic characteristics were diminished after the U.S.'s refit. However, by reason of the same cause and the blunter shell body tip and AP cap, horizontal armour-piercing capacity was ameliorated.

Post-war gunnery trials revealed that dispersion was approximately 2 % and 1.1 % of range at normal battle range before and after the introduction of firing delays, respectively.

Picture 1.2

The French battleship Richelieu, photographed on 18 May 1944.

Comparison of Main Batteries

Littorio and Richelieu carried the hardest hitting guns of this diameter ever put to sea. Both navies attained high performance by maximizing pressure and selected projectiles of virtually identical weight. Consequently, the guns shared the same strengths and weaknesses at large. These were outstanding ballistic characteristics and vertical armour penetration capacity on the one side, and low efficiency against horizontal armour, intensive bore erosion and dispersion on the other. Unrelated to pressure was the relatively low rate of fire, a shortcoming both guns shared.

The most noteworthy difference between the two weapons was that Littorio's guns had longer barrels and, having effectively identical average pressure as designed, proportionately greater muzzle energy, further increasing the strengths and weaknesses characterizing high-velocity guns. Attention is invited to the disparity between the explosive content and length of AP type projectiles and the high emphasis placed on underwater performance by the French. Also, the Italian AP projectiles had much larger and blunter armour-piercing caps than the domestic French shells, which probably ensured superior performance against very heavy armour, especially at high obliquities.

After the U.S.'s refit of Richelieu, the performance of the primary armament of the two ships no longer paralleled one another so closely. With their muzzle velocity decreased, ballistic characteristics of the French guns more appreciably fell behind those of their Italian counterparts, which resulted in diminished vertical armour penetration capacity and accuracy, but superior horizontal armour penetration capacity and durability. The size and shape of the AP cap of the Crucible Steel projectiles manufactured for Richelieu were more comparable to those of the Italian AP shells, but the tip of the shell body in the American design was blunter.

Table 1.1

Armament Comparison (as designed, new guns, firing AP projectiles)					
Designation	Littorio		Richelieu		
Gun)					
Muzzle Energy (MJ)	335		304		
Muzzle Velocity (fps/mps)	2,854	870	2,723	830	
Maximum Range (yds/m)	46,807	42,800	45,604	41,700	
Firing Cycle (sec)	45		45		
AP Projectile)					
Weight (lbs/kg)	1,951	884.8	1,949	884.0	
Bursting Charge (lbs/kg/%)	22.4	10.16	1.1	48.3	21.9	2.5
Length (in/mm/cal.)	66.9	1,700	4.47	74.8	1,900	5.0
Density Factor (lbs/in^3)	0.58		0.58		
Sectional Density (lbs/in^2)	8.67		8.71		

Table 1.2

Broadside Firepower Comparison (as designed, new guns, firing AP projectiles)				
Designation	Littorio		Richelieu	
Energy (MJ)	3,015		2,432	
Weight (lbs/kg)	17,559	7,963	15,591	7,072
Bursting Charge (lbs/kg)	202	91	386	175
Energy/min. (MJ)	4,020		3,243	
Weight/min. (lbs/kg)	23,412	10,618	20,788	9,429
Bursting Charge/min. (lbs/kg)	269	122	515	233

Table 1.3

Comparison of Ballistic Characteristics (I.V.: Littorio=2,789 fps/850 mps, Richelieu=2,723 fps/830 mps)								
Range	Angle of Descent		Danger Space (h=20 ft/6.1 m)		Striking Velocity		Terminal Energy	
yds (m)	deg.		yds (m)		fps (mps)		MJ	
---	Littorio	Richelieu	Littorio	Richelieu	Littorio	Richelieu	Littorio	Richelieu
10,936 (10,000)	5.0	5.2	76.2 (69.7)	73.3 (67.0)	2,254 (687)	2,215 (675)	209	201
16,404 (15,000)	8.7	9.0	43.6 (39.9)	42.1 (38.5)	2,034 (620)	1,995 (608)	170	163
21,872 (20,000)	13.4	14.0	28.0 (25.6)	26.7 (24.5)	1,847 (563)	1,814 (553)	140	135
27,340 (25,000)	19.3	20.2	19.0 (17.4)	18.1 (16.6)	1,719 (524)	1,686 (514)	121	117
32,808 (30,000)	26.1	27.2	13.6 (12.5)	13.0 (11.9)	1,634 (498)	1,608 (490)	110	106
38,276 (35,000)	37.6	38.1	8.7 (7.9)	8.5 (7.8)	1,585 (483)	1,572 (479)	103	101

Note= According to secondary sources, Richelieu's striking velocity at 21,872 yds (20,000 m) was 1,785 fps (544 mps), but this value cannot possibly be correct. The value given here (1,814 fps/553 mps) is the result of interpolation.

The disparity between the angle of descent of the two projectiles at 38,276 yds (35,000 m) is inconsistent with the tendency prevailing up to 32,808 yds (30,000 m), which hints that these values may not be entirely precise, though it is unlikely that they are significantly off.

Table 1.4

Armour Penetration Capacity Comparison Based on Bu. Ord. Sk. 78841 – in (mm) (I.V.: Littorio=2,789 fps/850 mps, Richelieu=2,723 fps/830 mps)						
Range	Vertical Armour at 90 deg.		Vertical Armour at 60 deg.		Horizontal Armour	
yds (m)	Littorio	Richelieu	Littorio	Richelieu	Littorio	Richelieu
10,936 (10,000)	25.10 (638)	24.62 (625)	19.11 (485)	18.76 (477)	1.40 (36)	1.44 (37)
16,404 (15,000)	21.91 (557)	21.41 (544)	16.82 (427)	16.46 (418)	2.31 (59)	2.35 (60)
21,872 (20,000)	18.90 (480)	18.43 (468)	14.68 (373)	14.34 (364)	3.26 (83)	3.34 (85)
27,340 (25,000)	16.35 (415)	15.83 (402)	12.87 (327)	12.49 (317)	4.36 (111)	4.47 (114)
32,808 (30,000)	14.08 (358)	13.61 (346)	11.26 (286)	10.91 (277)	5.59 (142)	5.74 (146)
38,276 (35,000)	11.16 (283)	10.95 (278)	9.15 (232)	8.99 (228)	7.88 (200)	7.93 (201)

Figure 1.1

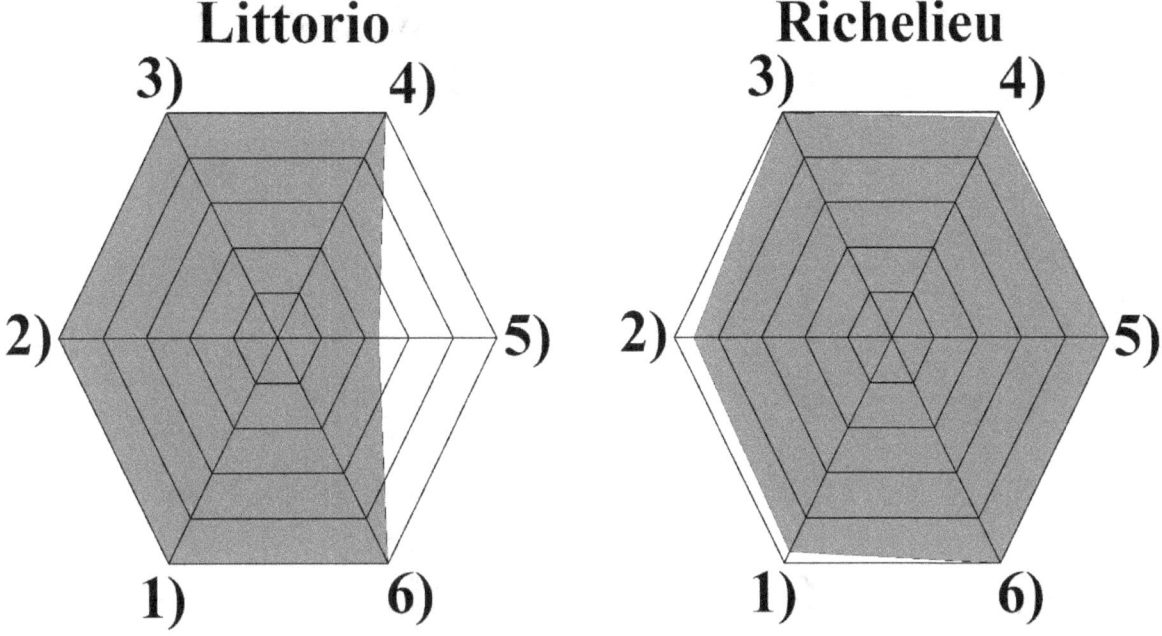

Figure 1.2

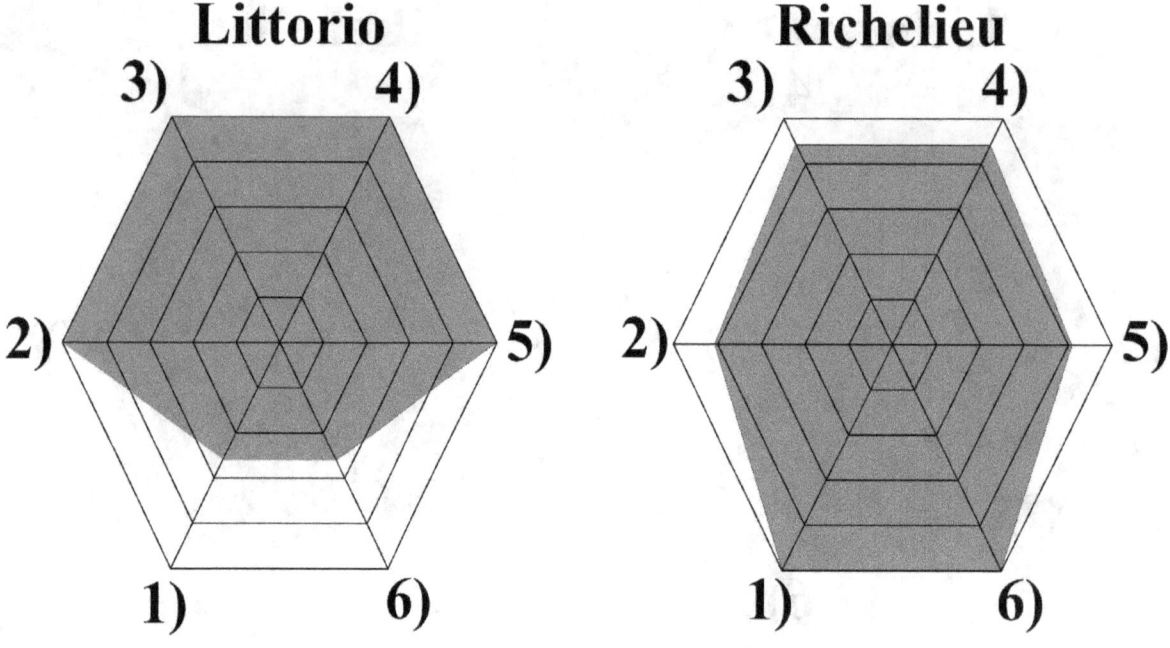

1) Bursting Charge
2) Energy
3) Weight
4) Weight/min.
5) Energy/min.
6) Bursting Charge/min.

Figure 1.3

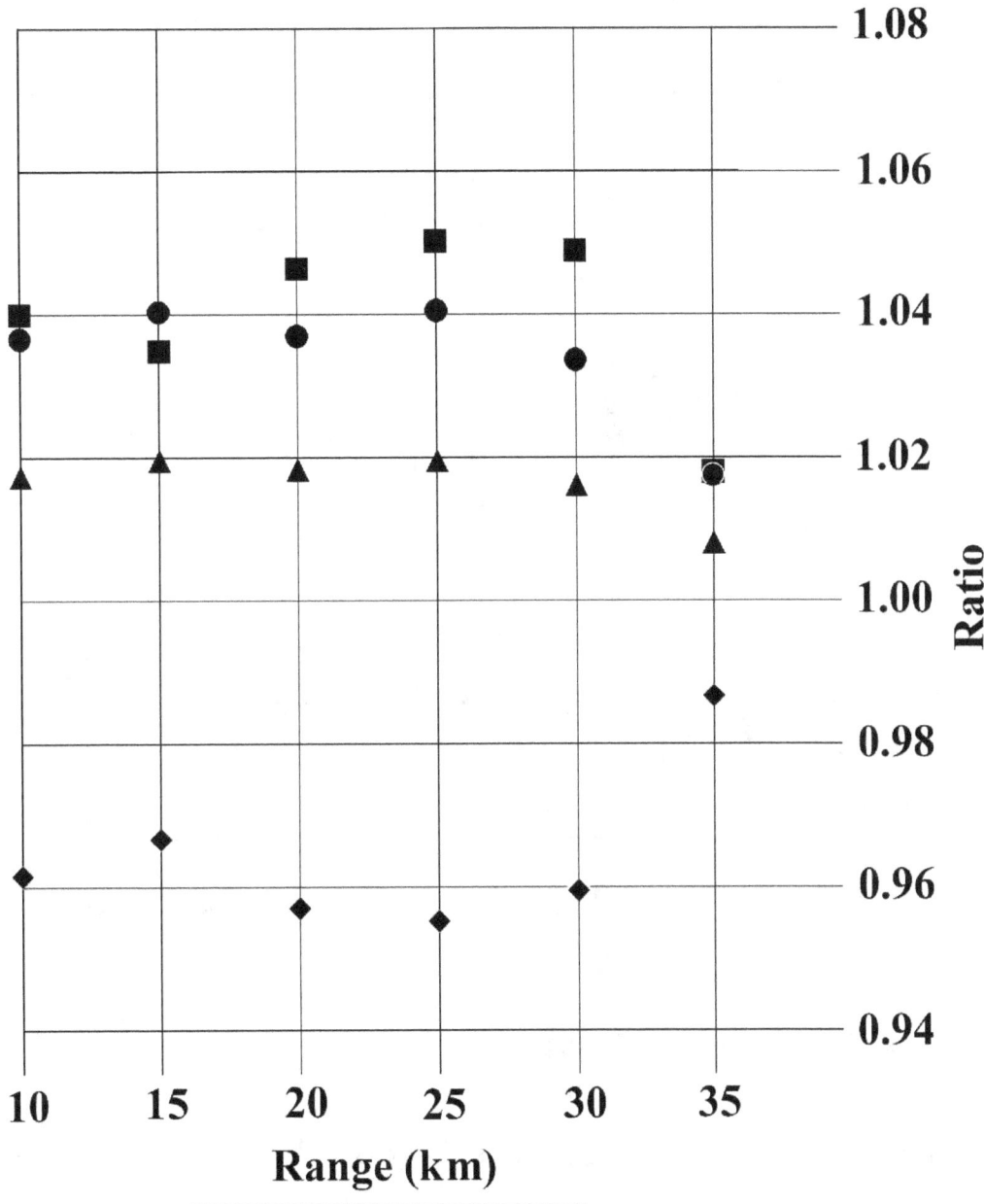

Figure 1.4

Armour Penetration Capacity of Littorio's Main Guns

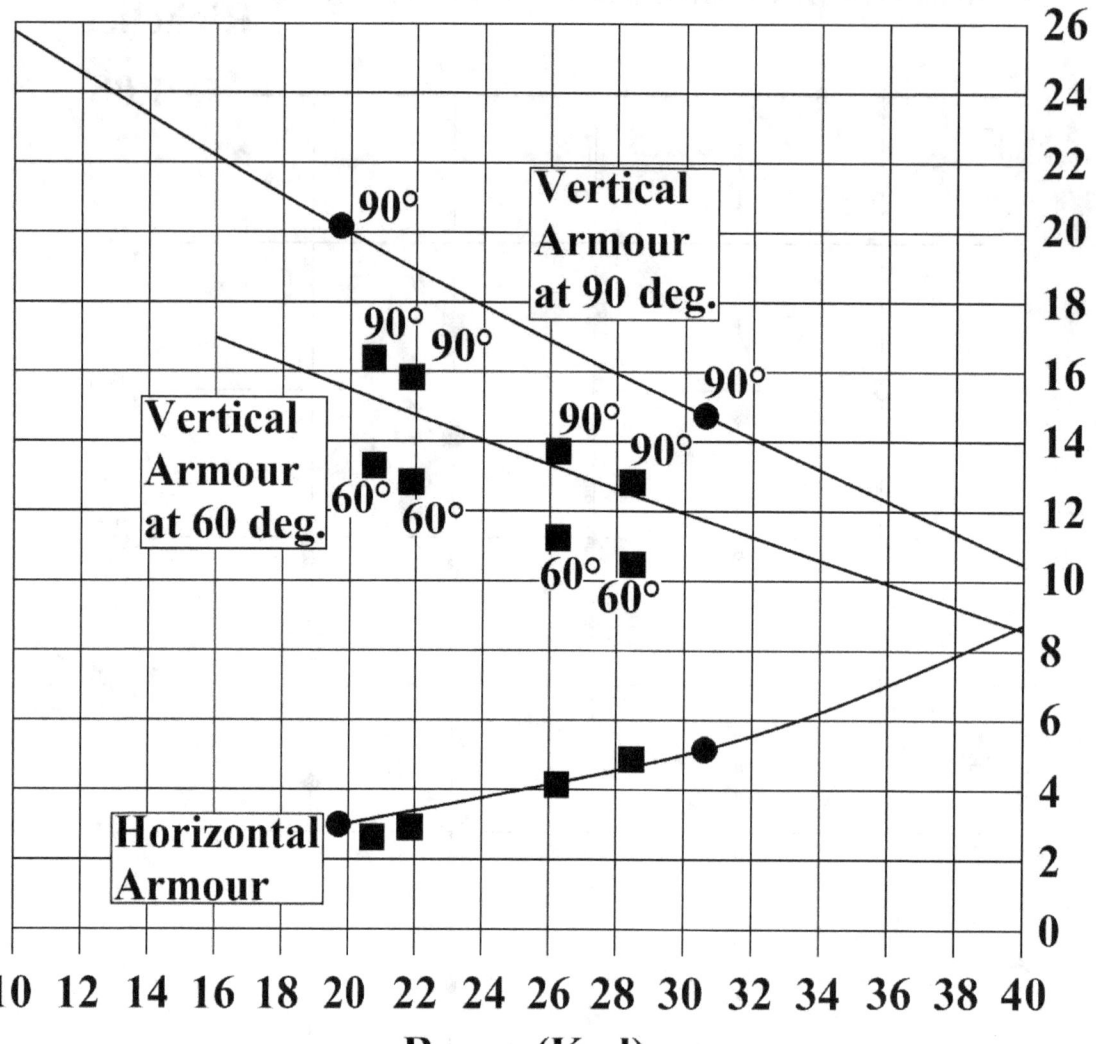

Curves=Bu. Ord. Sk. 78841
● = O.T.O. Melara
■ = Direttive e norme per l'impiego della Squadra navale. Impiego delle artiglierie, September 1942.

Description of Figure 1.4:

According to the Italians, the values dated September, 1942 are pessimistic, although the horizontal armour piercing values are in reasonably good agreement with those supplied by O.T.O. Melara and Bu. Ord. Sk. 78841. The vertical armour piercing values are apparently based on the DeMarre formula:

$$t=[v*\cos(\Theta)*m^{1/2}/(K*d^{3/4})]^{10/7}$$

Whence
t=thickness {mm},
v=velocity {mps},
m=mass {kg},
d=diameter {mm},
Θ=obliquity {deg.},
K=coefficient {2.87}.

The values given by O.T.O. Melara are in complete agreement with Bu. Ord. Sk. 78841. According to the Italians, the O.T.O. Melara values should be used when comparing the performance of new guns with those of foreign ones, indicating that Bu. Ord. Sk. 78841 is a good approximation too.

Figure 1.5

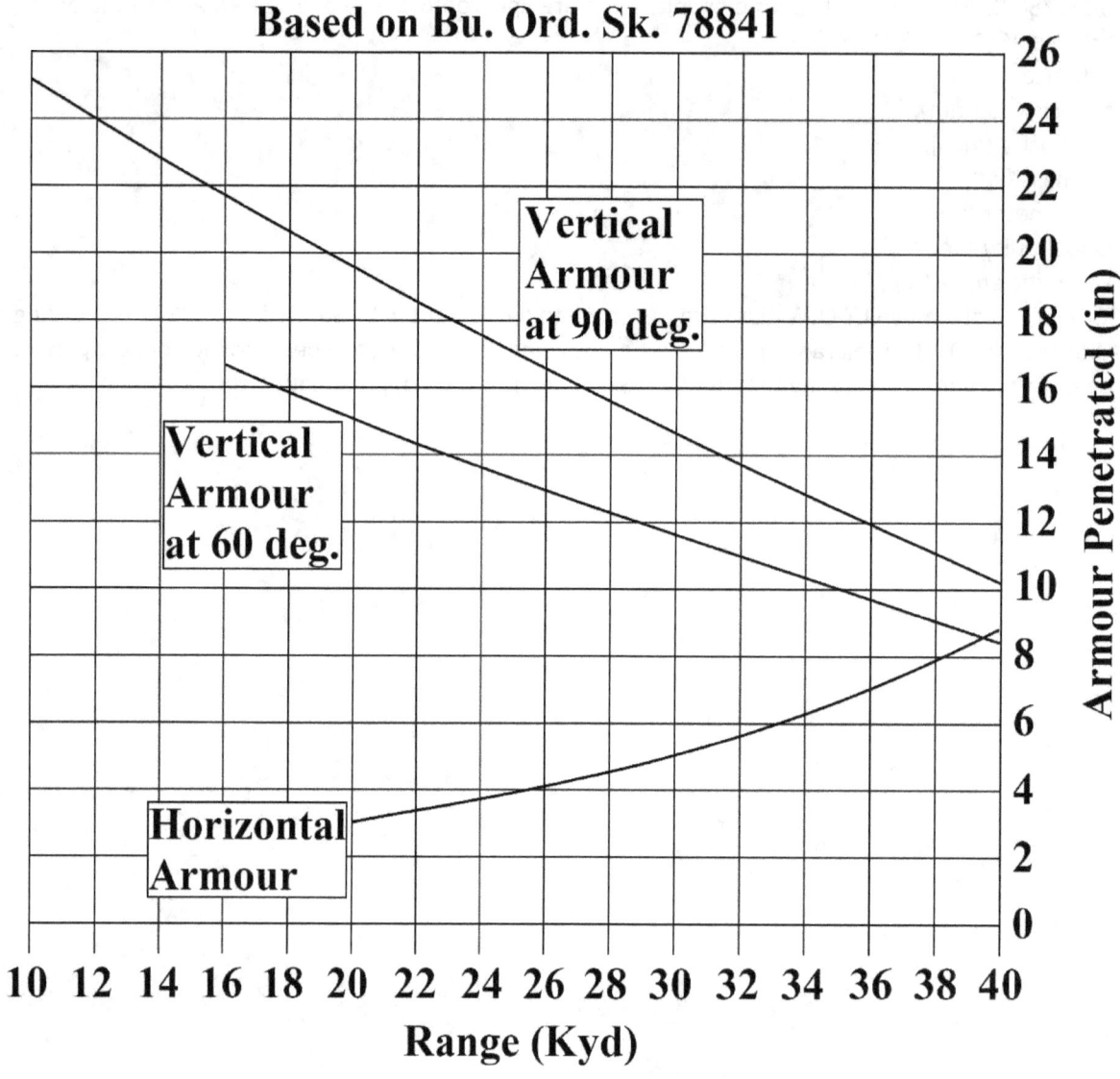

Figure 1.6

Armour Penetration Capacity Comparison
Based on Bu. Ord. Sk. 78841
R
L
Vertical Armour at 90 deg.
L
R
Vertical Armour at 60 deg.
R
L
Horizontal Armour
26
24
22
20
18
16
14
12
10
8
6
4
2
0
Armour Penetrated (in)
10 12 14 16 18 20 22 24 26 28 30 32 34 36 38 40
Range (Kyd)
L=Littorio
R=Richelieu

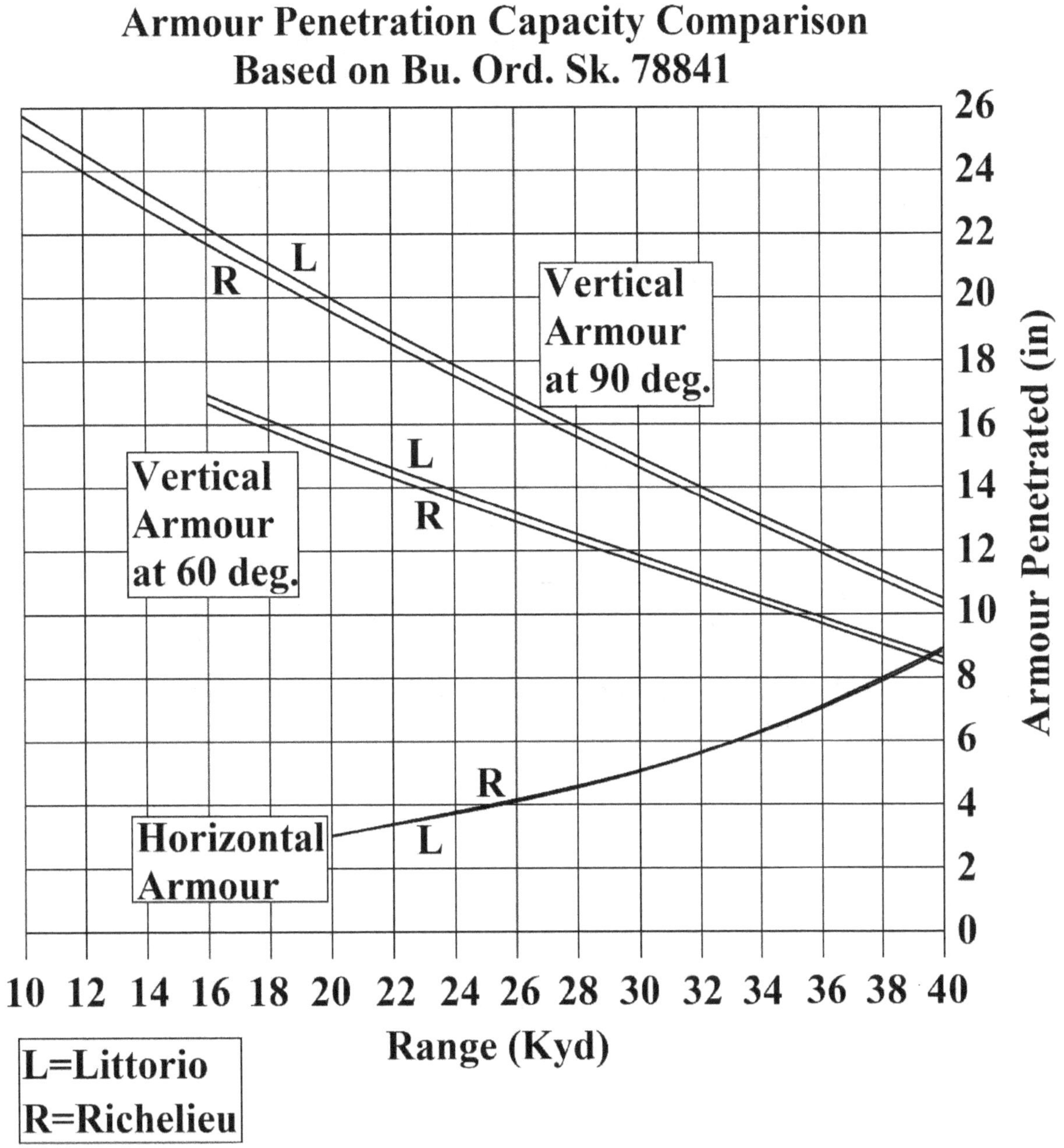

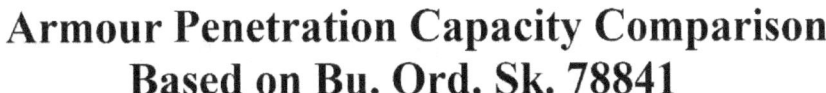

Medium Caliber Guns

Table 1.5

Secondary Armament Comparison (HE/SAP Projectiles)				
Designation	Littorio		Richelieu	
Muzzle Energy (MJ)	22.0		21.6	
Muzzle Velocity (fps/mps)	3,264	995	2,854	870
Maximum Range (yds/m)	27,231	24,900	28,981	26,500
Rate of Fire (Ro./min.)	4-5		5-6	
Projectile Weight (lbs/kg)	97.9	44.4	125.9	57.1
Bursting Charge (lbs/kg/%)	5.2	2.3 / 5.3	4.2	1.9 / 3.3

Table 1.6

Secondary Gun Broadside Firepower Comparison				
Designation	Littorio		Richelieu	
Energy (MJ)	132		130	
Weight (lbs/kg)	582	266	755	343
Bursting Charge (lbs/kg)	30	14	25	11
Energy/min. (MJ)	594		713	
Weight/min. (lbs/kg)	2,619	1,197	4,154	1,884
Bursting Charge/min. (lbs/kg)	137	62	138	63

As the tables indicate, the secondary battery of both ships was reasonably well-adapted to harassing fire, but the French vessel had an edge.

Chapter 2 – Fire-Control

Littorio's Main Battery Fire-Control

Littorio's main battery director was located atop the fire-control tower of the battleship. The director incorporated a 23.62' (7.20 m) duplex rangefinder consisting of one stereoscopic and one coincidence type optical system. The lower turret was equipped with a stereoscopic rangefinder of the same base length, but this was not used for gunnery.

All three of the ship's main gun turrets housed one 39.37' (12.0 m) coincidence rangefinder and one stereoscopic rangefinder of identical base length in duplex installations.

During the early 1940s, EC 3/ter "Gufo" radar instruments were mounted on the fire-control tower of all three Littorio class battleships. These had the following specifications:

Table 2.1

Frequency	400 Mhz
Peak Power	10 kW
Pulse Repetition Frequency	500 Hz
Pulse Duration	4 Milliseconds
Wavelength	29.5in (75 cm)
Horizontal Beam Width	6 deg.
Vertical Beam Width	20 deg.
Range (Battleship)	32.8 kyd (30 km)
Range (Aircraft)	87.5 kyd (80 km)
Display	'J'
Antenna	'Horns'

The instrument had various teething problems and was not adopted for fire-control.

For illuminating surface targets at night, the ship carried four 59.1" (150 cm) and two 41.3" (105 cm) searchlights. Furthermore, starshells were developed for both the 4.72" (120 mm) and 3.54" (90 mm) guns.

The ship's air group initially consisted of three Ro. 43 floatplanes, but from 1942 Re. 2000 type fighters could also be carried. The ship's air group from this point on was a combination of floatplanes and fighters.

Picture 2.1

The main battery director of the Italian battleship Roma is turned to port. (1940)

Richelieu's Main Battery Fire-Control

Richelieu had two, Jean Bart one, primary armament directors. The forward director (Télépointeur A) of both ships was equipped with an OPL 45.93' (14.0 m) triplex stereoscopic rangefinder, while the aft director (Télépointeur B) of Richelieu incorporated an OPL 26.25' (8.0 m) duplex stereoscopic rangefinder.

Both of Richelieu's main turrets were fitted with an OPL 45.93' (14.0 m) duplex stereoscopic rangefinder, those of her sister with OPL 46.59' (14.2 m) quadruplex stereoscopic rangefinders.

Richelieu received a prototype radar equipment in 1941 with the following specifications:

Table 2.2

Designation	DEM
Range (Aircraft at 4,921 ft/1,500 m)	87.5 kyd (80 km)
Range (Aircraft at 3,281 ft /1,000 m)	54.7 kyd (50 km)
Range (Battleship)	21.9 kyd (20 km)
Range Accuracy	547 yd (500 m)

While this was a rudimentary instrument, in 1944-45 the ship acquired more advanced sets including British Type 281B (air search), Type 284P (main gun fire-control), Type 285P (secondary gun fire-control), American SA-2 (air search), SF (surface search) and SG-1 (surface search) radars. Of particular interest to us is the British Type 284P main battery gunnery radar with which the forward director was equipped in 1944:

Table 2.3

Designation	Type 284P
Range (Battleship)	24.1 kyd (22.0 km)
Range Accuracy	27 yd (25 m)
Bearing Accuracy	0.2°
Range Resolution	164 yd (150 m)
Bearing Resolution	1°
Power	150 kW
Wavelength	50 cm
Frequency	600 MHz

As completed, Jean Bart carried DRBV 11 (air/surface search), DRBV 20 (air search), DRBV 30 (navigation), DRBC 10A, (fire-control), DRBC 30B (fire-control) and ACAE (fire-control) radars.

Richelieu had five remotely controlled 47.2" (120 cm) searchlights. Starshells were designed for both the 3.94" (100 mm) and 6" (152.4 mm) guns, but it is unclear whether the larger caliber was ever manufactured.

The ship's designed air group consisted of four Loire 130 reconnaissance/spotting seaplanes, although she carried no more than three in practice. Facilities for these aircraft included a hangar capable of accommodating two seaplanes with folded wings, two catapults situated abaft of the hangar arranged *en echelon*, aviation gasoline fuel tanks and a seaplane hoist with a maximum capacity of 4.5 m.t. Along with the entirety of the ship's obsolete close-range anti-aircraft battery, aviation facilities were suppressed in January 1943 as the more numerous (and powerful) anti-aircraft artillery proposed by the Americans required substantial deck space.

Picture 2.2

The French battleship Richelieu, after refit in the U.S., in September 1943. Note that the foretop secondary armament director had been removed.

Comparison of Main Battery Fire-Control Systems

The base length of the two ships' turret rangefinders was comparable with the French having a slight edge. However, owing to their low position, it may not be possible to use turret rangefinders at very long ranges. Under such conditions, Richelieu had a categorical advantage, having one additional elevated artillery position and her foretop director rangefinder being almost twice as large and incorporating one additional optical system.

While neither navy had effective domestic fire-control radars, there can be no doubt that after Richelieu's late-war refits, the French enjoyed a distinct advantage from this perspective.

Unfortunately, we do not possess enough empirical data to evaluate the performance of the below-deck fire-control equipment of the two ships, which makes it exceedingly difficult to draw a meaningful conclusion regarding the overall performance of their fire-control systems.

Table 2.4

Fire-Control Equipment Comparison		
Main Battery Rangefinders – Base Length in ft (m)/Type)		
Designation	Littorio	Richelieu
Foretop Director	23.62 (7.2) Duplex/1C+1S	45.93 (14.0) Triplex/3S
Aft Director	---	26.25 (8.0) Duplex/2S
Turret No. I.	39.37 (12.0) Duplex/1C+1S	45.93 (14.0) Duplex/2S
Turret No. II.	39.37 (12.0) Duplex/1C+1S	45.93 (14.0) Duplex/2S
Turret No. III.	39.37 (12.0) Duplex/1C+1S	---
Radar (domestic))		
Equipment	EC 3/ter 'Gufo'	DEM
Range (Battleship)	32.8 kyd (30 km)	21.9 kyd (20 km)
Other Equipment)		
Searchlights	4 x 59.1 in (150 cm) + 2 x 41.3 in (105 cm)	5 x 47.2 in (120 cm)
Aircraft	1-3 x Ro. 43 + 2-0 x Re. 2000	0-4 × Loire 130

Figure 2.1

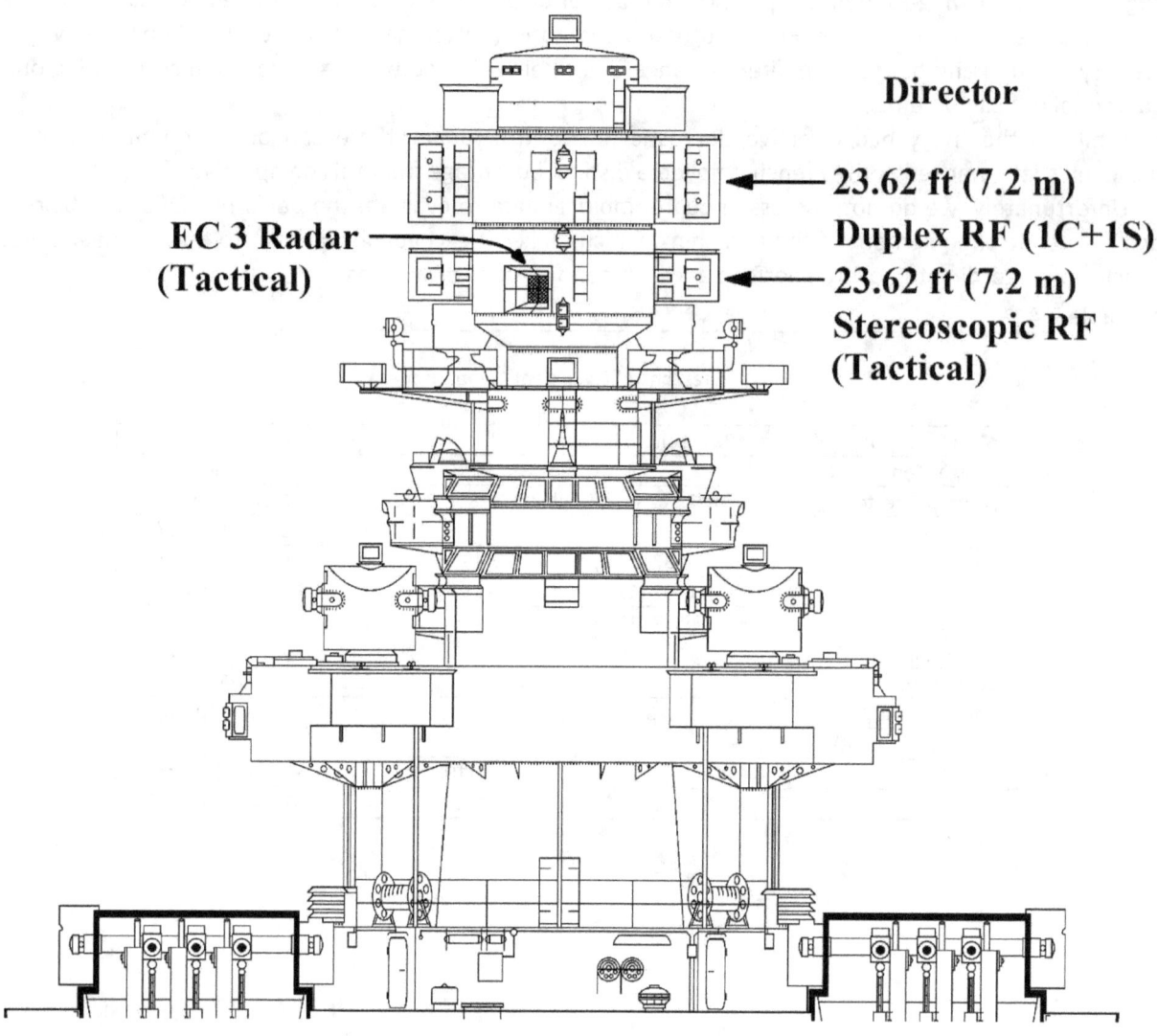

Ro. 43 Aircraft

Main Turret

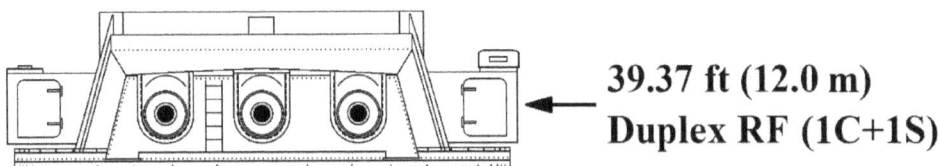

← 39.37 ft (12.0 m)
Duplex RF (1C+1S)

Figure 2.2

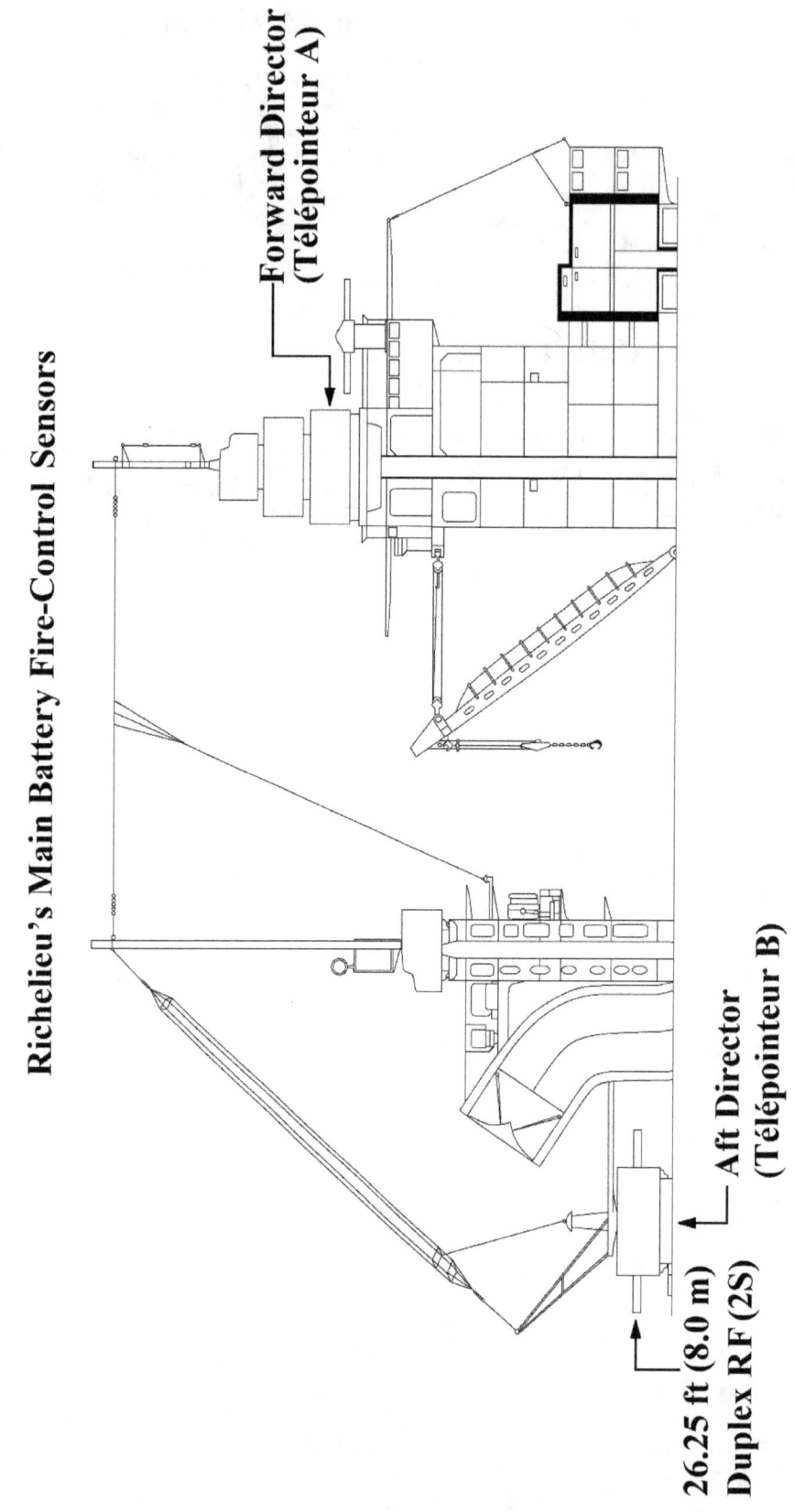

Richelieu's Main Battery Fire-Control Sensors

Forward Director (Télépointeur A) — As Designed

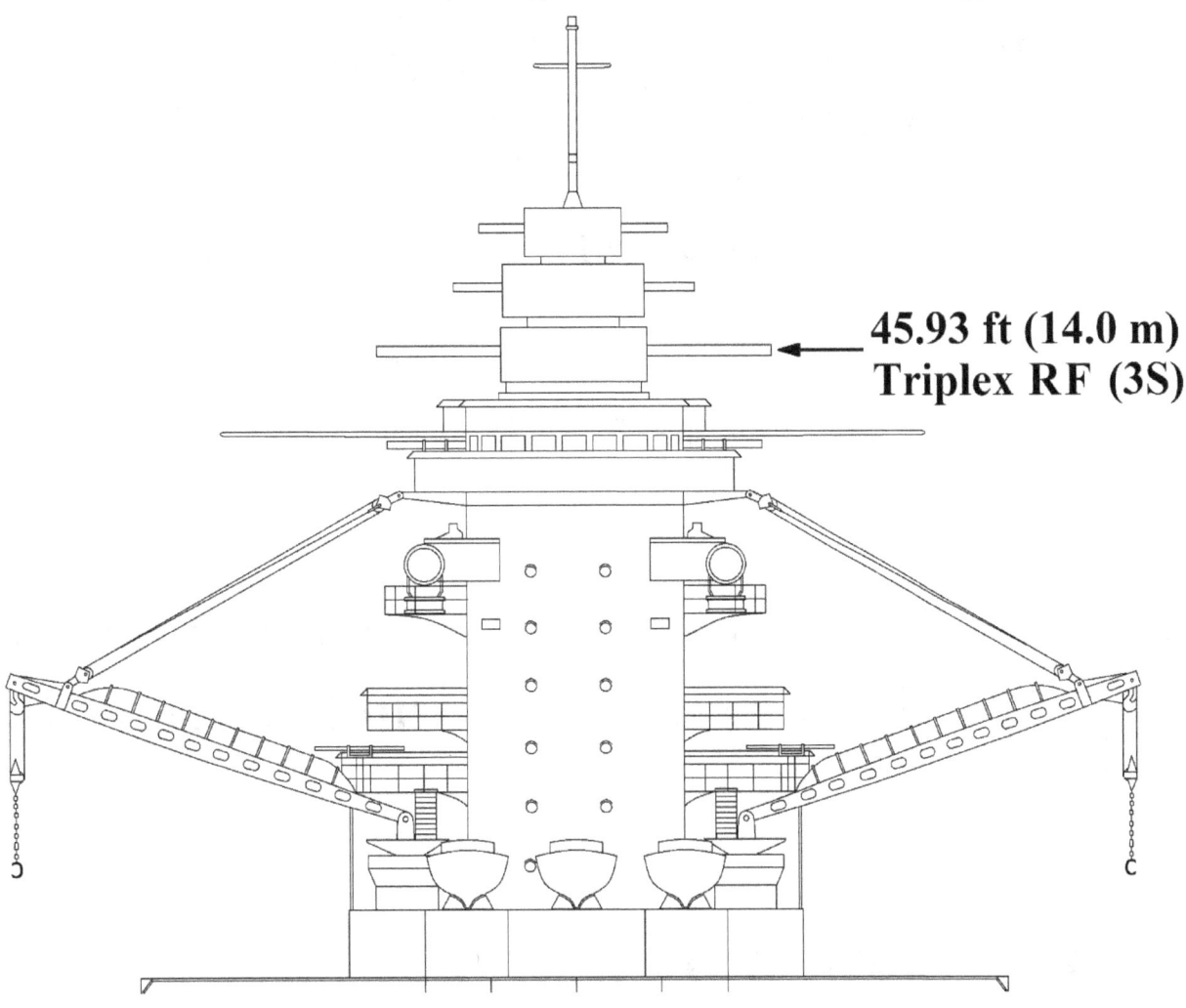

← 45.93 ft (14.0 m) Triplex RF (3S)

Forward Director (Télépointeur A) — Refitted

45.93 ft (14.0 m) Triplex RF (3S)

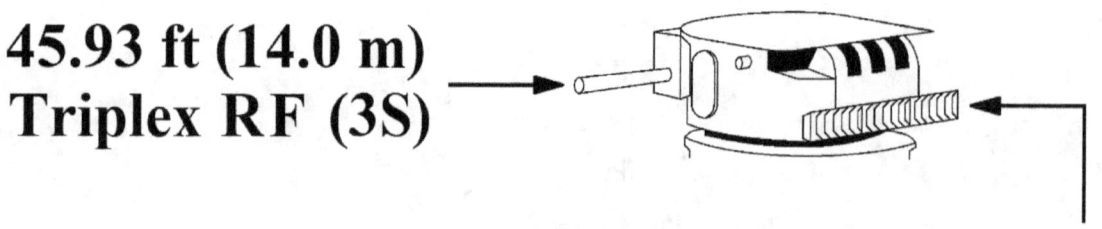

Type 284P Fire-Control Radar

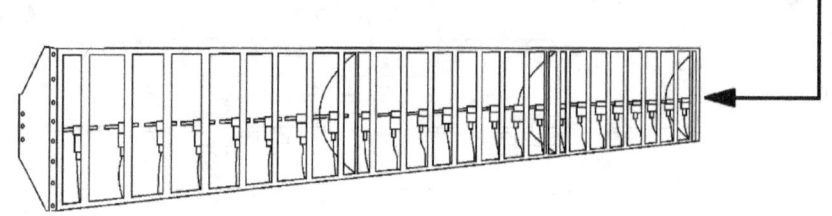

Main Turret

45.93 ft (14.0 m) Duplex RF (2S)

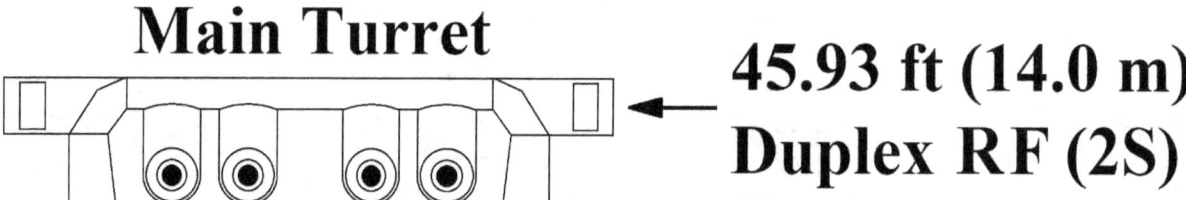

Lorie 43 Aircraft

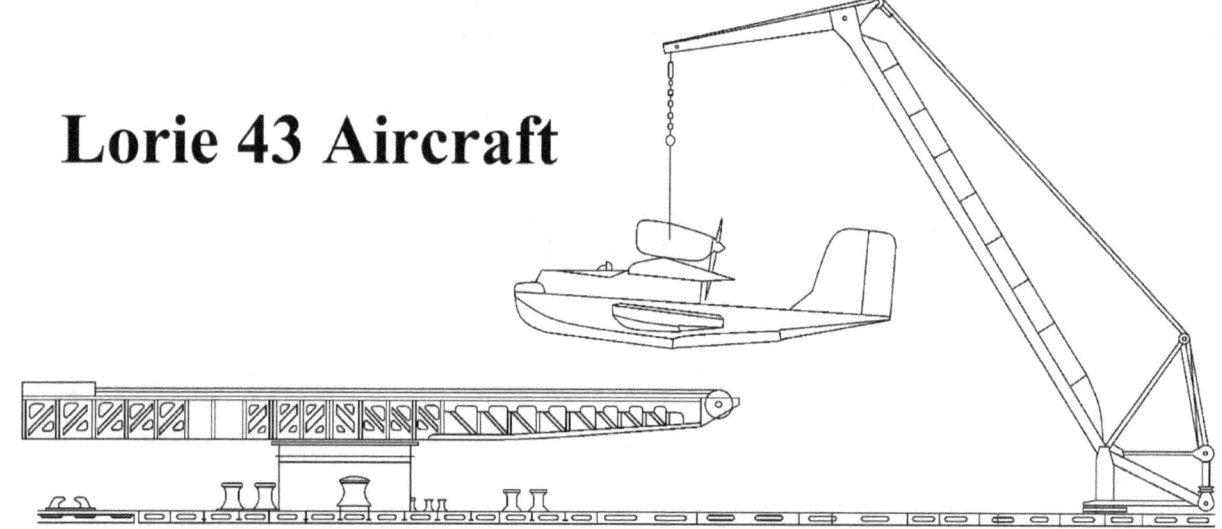

Chapter 3 – Protection

Littorio's Protection Scheme

Even though weight reserved for protection was percentagewise unremarkable, this was, to some extent, compensated by compressing the citadel and inclining the side armour of the ship from the vertical. However, the Italians sacrificed notable weight to provide medium armour protection outside the citadel and heavy armour plating to the medium caliber mounts.

The vertical armour of these ships was unique in that it was a spaced assembly composed of an outer plate of medium thickness designed to decap APC projectiles before they could make contact with the main belt, which was followed by two inclined, longitudinal, splinter bulkheads and the upper strake of the innermost holding bulkhead. Based on ADM 281/31, the limit velocity of such spaced assemblies is comparable to that of a single plate of the same total poundage, but they have the advantages of more reliably breaking up projectiles and simplifying the manufacturing of armour plates. The Italians estimated that the vertical armour of the citadel was sufficient to secure the ship against their newly developed, high-performance 15" (381 mm)/50 guns beyond 17.5 kyd (16.0 km), albeit, based on our calculations, this value corresponds to 60 deg., not 90 deg., target angle.

The Italians failed to reserve enough weight for a deep armour belt, however. As a result, the ship had limited ballistic underwater protection at light displacement and marginal armoured freeboard at deep load. The latter was remedied by increasing the thickness of the upper side hull and the upper deck to secure the compartments above the armour deck against at least HE and light AP type shells and shrapnel. This system also initiated the fuze mechanism of bombs and decapped APC projectiles before making contact with the main armour deck.

Although the main protective deck above the magazines was very heavy at the center, its thickness at the sides and atop the machinery compartments was below Allied and Japanese standards. Still, plate thickness was adequate to keep out high-velocity projectiles up to extreme ranges, but these ships would have been vulnerable against the largest low-velocity guns of the Pacific Theatre and the Royal Navy at long range, including their magazines, which could be reached by artillery fire through the thinner sides of the armour deck. While it is true that more effective, and reliable, protection against plunging shellfire could have been attained by an armour deck of uniform thickness, it should be kept in mind that the Italians probably considered aerial AP bombs a more imminent threat than plunging shellfire, and these, due to their almost vertical descent, could not, in theory, reach the magazines without having to defeat the thickest plates.

The ship's torpedo defense system was built around a Pugliese-type shock absorbing cylinder. This was a tube with a maximum diameter of 12.5' (3.8 m) at midship, running parallel to the outer skin inside the underwater hull of the ship. There was a light curved bulkhead between the outer skin and the cylinder. The main torpedo bulkhead, which was also curved, was behind the cylinder. The compartment between the curved bulkheads and the cylinder was liquid loaded, but the cylinder itself was empty. It was envisaged that the energy of underwater detonations would be largely expended by crushing the tube. The system was enclosed by a light, vertical, longitudinal bulkhead raised behind the main torpedo bulkhead.

The citadel of the ship was terminated by vertical transverse bulkheads, although their tapered lower strake was inclined from the vertical. The transverse bulkheads were slightly lighter than in modern foreign ships of this size, but they were shielded by medium armour carried outside the citadel.

In terms of poundage, the vertical armour of the main turrets and turret barbettes was similar to that of the citadel, but they had inferior effective resistance due to either less favorable inclination (turret face) or the lack of it (barbettes). Unlike the citadel, they also lacked the benefits associable with spaced armour. Consequently, the main armament of the ship had only marginal immunity against modern 15" (381 mm) and larger guns, notwithstanding that their horizontal protection was quite satisfactory.

The conning position had comparatively light armour protection and no immunity against battleship guns. The superstructure beneath the conning tower was protected by medium armour.

The uptakes were secured by one deck level high, relatively heavily armoured cylinders. It is perhaps worthwhile to note that foreign navies disowned this design trait by the time of the last shipbuilding race and combine light splinter bulkheads with heavy armour grating to protect the uptakes of their new ships instead; HMS Rodney was the newest foreign battleship carrying a similar system.

The ship had two steering gear compartments, both of which, along with the shaft tunnels, were protected by a heavy armour deck with sloped sides. Although these slopes could be perforated by major caliber attack regardless of battle range, the steering ability of all modern capital ships, with the exception of those commissioned by the U.S. and Japan, had similar protection. The auxiliary steering gear compartment was limited by a medium, the primary by a heavy, transverse bulkhead.

The bow section of the ship was reinforced by a splinter belt of medium thickness. It was topped by an armour deck of appropriate poundage. This system was enclosed by an armoured transverse bulkhead circa halfway between the forward transverse bulkhead and the extremity of the bow, slightly closer to the former.

The ship's heavy vertical armour was constructed of KC type armour manufactured by Terni. Vertical armour of medium thickness and the ship's armour decks were made of OD. ER, an armour grade construction steel, served as the backing of heavier armour plates. Splinter bulkheads and the torpedo defense system of the ship were also made of this material.

Based on the armour penetration capacity of Littorio's guns as described by O.T.O. and the designed inner limit of immunity of the ship's citadel, the ballistic efficiency of Italian armour closely approximated Bu. Ord. Sk. 78841 specifications, assuming that the Italian AP projectiles had appropriate mechanical properties to simulate the piercing capacity of U.S. shells.

Picture 3.1

The Italian battleship Vittorio Veneto, photographed in 1940 before completion. Notice the external side armour of the battleship's citadel. (Compare the armoured freeboard of the ship with that of the completed Roma – picture 2.1. Note that the tapering upper segment is only a covering plate, not armour.)

Richelieu's Protection Scheme

The designers of the new battleship class devoted a high percentage of its displacement to protection and preferred a weight saving approach to armour distribution. The armament, horizontal protection and the citadel as a whole were very concentrated, while side armour was inclined from the vertical. The designers sacrificed very little armour to protect the hull outside the citadel, providing only a light platform deck to reinforce the bow section and a turtle deck and medium transverse bulkheads to shield the shaft tunnels and the steering gear machinery, but no side armour. The secondary armament was also left vulnerable against guns larger than 6" (152 mm) in diameter. However, these weight-saving methods allowed the designers to provide very heavy ballistic protection to the vitals.

The main protective deck was exceptionally heavy, especially atop the forward magazines. There was a light second armour deck, or splinter deck, beneath the main armour deck. The sides of this deck were inclined from the horizontal.

The upper edge of the main armour belt was connected to the main, its lower edge to the sloped sides of the lower, armour deck. This system, which the French named *caisson blindé* (armoured box), provided a multi-layer defense in the form of an outer layer of heavy ballistic protection formed by the main deck and main belt, followed by an internal layer of splinter barriers formed by the lower armour deck and its sloped sides.

The nethermost strake of the main belt was tapered; this was justified by the greater drag force of water as compared to air. The armour belt was deeper than in Italian and German designs, although it was not as deep as in contemporary Allied and Japanese ships.

The armour belt was inside the hull, but its upper edge touched the outer skin. The longitudinal compartment between the main belt, the outermost holding bulkhead and the hull was filled by a water repellent material termed *ebonité mousse* (ebonite foam).

The underwater protection system was formed by a series of vertical, longitudinal, holding bulkheads beneath the lower armour deck. A total of five bulkheads separated the system into five longitudinal compartments. As already described, the outermost compartment contained water-excluding foam. The second and fourth compartments were kept void, the third was liquid loaded while the fifth was a cable tunnel. The fourth bulkhead was the main holding bulkhead. It was integrated into the vertical protection of the citadel and provided ballistic protection against underwater hits. An extension of this bulkhead served as a splinter barrier between the two armour decks, but this strake had diminished thickness. In order to compensate for the reduction of hull volume – and consequently system depth – nearing the limits of the citadel, the thickness of the main holding bulkheads was increased in this region.

The volume of the torpedo defense system of Jean Bart was augmented as compared to that of Richelieu by blistering her underwater hull abreast the citadel.

The citadel was terminated by vertical transverse bulkheads. The thickness of these bulkheads varied depending on whether they were reinforced by other armour plates or not. The forward transverse bulkhead was thickest between the lower armour deck and the forward platform deck, but its thickness was reduced above the former and below the latter. Thickness was at its minimum outside the main holding bulkheads. The thickness of the aft transverse bulkhead was uniform between the main holding bulkheads, but was reduced outboard.

In order to protect them from mines, the magazines were reinforced by lightly armoured bottom decks.

The uptakes were secured by armour grating twice the thickness of the main deck and light splinter bulkheads.

The French were greatly concerned about the risk of losing the entirety of a quadruple turret unit in action, as it amounted to half of the main battery firepower of the ship. In order to minimize this threat, the turrets were internally subdivided by longitudinal bulkheads of medium thickness. Also, the turrets were placed as far apart from each other as possible so as to diminish the probability of a single hit damaging both.

The gunhouses and barbettes had heavier armour plating than in any other contemporary European design. Still, the effective resistance of the main armament did not match that of the citadel. This was due to the unfavorable inclination of the face and fore roof plates.

The conning position had reasonably heavy armour protection, but it had no reliable immunity against modern major caliber guns.

The directors and fire-control towers had bulletproof plating.

Heavy vertical armour and the roof plate of the main turrets and that of the conning position were made of cemented, the armour decks and the rangefinder and sight hoods of the turrets of non-cemented, armour. The floor, communication shaft and gastight doors of the conning tower and the main holding bulkheads were constructed of special steel. The glacis of the main turrets and the lower ring bulkheads were made of high tensile, the bulletproof plating of the directors and fire-control towers of hardened, steel.

Little is known about the ballistic efficiency of heavy French armour. The only empirical data at our disposal are the damage reports describing the effects of the large-caliber hits French capital ships sustained during the war. A number of these shells hit unarmored parts, but five hit the armoured citadel of Jean Bart, one that of Richelieu and three that of Dunkerque. However, only one of these hits has any representative value regarding the quality of French armour, as the ballistic limit velocity of the target in the other instances was either far-above or well-below the striking velocity of the projectile and, as might be expected, the projectile ricocheted and attained penetration in the former and in the latter cases respectively. The only hit that has representative value was the first hit scored by USS Massachusetts against Jean Bart. At a range of circa 24 kyd (21.9 km), Massachusetts penetrated the main armour deck and splinter deck of Jean Bart. Now, the angle of descent of Massachusetts' shells was circa 25 deg. at this range. Based on Bu. Ord. Sk. 78841, the combined limit velocity of the two decks was about 1,710 fps (521 mps) if hit at 65 deg. obliquity. However, the striking velocity of the shell was only about 1,480 fps (451 mps). This indicates that the mean limit velocity of these plates was less than 87 as a percentage of Bu. Ord. Sk. 78841. This is lower than the limit velocity of any of the German, American or Japanese homogeneous armour plates of comparable thickness described by NPG 5-47 and ATC 22/7/1948 that were tested in America. Notwithstanding that it would be unwise to jump at far-fetched conclusions based on a single impact, this is hardly a flattering result.

Picture 3.2

Richelieu en route to New York in 1943. Note the smooth hull of the battleship, revealing no sign of her internal side armour. This picture also shows that the inclination of the turret face plates from the vertical was no more than 20 deg., not 30 deg. as often cited.

Comparison of Protection Schemes

Though the vertical protection system of the two ships was fundamentally different, our calculations indicate that their effective resistance was similar, with Littorio's having a slight edge in terms of limit velocity and in that her system was more likely to break up projectiles based on empirical data.

However, Richelieu carried palpably more armour as compared to Littorio, and the Italian designers were also somewhat more disinclined to concentrate armour. This resulted in the French being able to provide markedly heavier horizontal armour, better protected main battery and conning position, as well as deeper side armour, to their new ships. Admittedly, even with heavier armour plating, Richelieu's main armament and conning position were not satisfactorily secured against high-performance 15" (381 mm) guns. The outer limit of the immunity zone of the French ship, on the other hand, exceeded that of Littorio, even if we assume that, as indicated by the performance of Jean Bart's deck at Casablanca against 16" (406 mm) attack, French homogeneous armour was of inferior quality.

Table 3.1

Littorio – Armour Protection			
Designation	Thickness		Material
	in	mm	
Citadel)			
Upper Belt	2.76+0.35	70+9	OD+ER
Outer Belt (15°)	2.76+0.35	70+9	OD+ER
Main Belt (15°)	11.02+0.47	280+12	KC+ER
Splinter Bulkhead No. I. (15°)	1.42	36	OD
Splinter Bulkhead No. II. (26°)	0.94	24	OD
Upper Deck	1.42+0.35	36+9	OD+ER
Second Deck	0.47	12	ER
Main Deck – Machinery/Sides	3.54+0.47	90+12	OD+ER
Main Deck – Machinery/Center	3.94+0.47	100+12	OD+ER
Main Deck – Magazines/Sides	3.94+0.47	100+12	OD+ER
Main Deck – Magazines/Center	5.91+0.47	150+12	OD+ER
Forward Transverse Bulkhead	2.76-8.27-3.94	70-210-100	---
Aft Transverse Bulkhead	2.76-8.27-2.76	70-210-70	---
Main Torpedo Bulkhead	1.10-1.57	28-40	ER
Main Turrets)			
Face Plate (30°)	14.96	380	KC
Fore Side Plate	7.87	200	KC
Rear Side Plate	5.12	130	KC
Rear	13.78	350	KC
Fore Roof Plate (7°)	7.87	200	KC
Rear Roof Plate	5.91	150	KC
Glacis	7.87	200	---
Overhang	3.94	100	---
Barbettes)			
Upper	13.78	350	KC
Lower – Turret No. III.	11.42	290	KC
Lower – Turrets No. I. and II.	11.02	280	KC
Conning Tower)			
Forward	9.84+0.39-10.04+0.98	250+10-255+25	KC+ER
Aft	6.89+0.98-7.87+0.39	175+25-200+10	KC+ER
Roof	3.54+0.39-4.72+0.39	90+10-120+10	OD+ER
Deck	3.54+0.39	90+10	OD+ER
Tube	7.87	200	OD
Superstructure Below CT	2.36	60	OD
Steering Gear Room, Shaft Tunnels, Stern Section)			
Upper Deck	0.47	12	ER
Second Deck	1.42+0.31	36+8	OD+ER
Third Deck	3.54-3.94+0.31	90-100+8	OD+ER
Aftermost Bulkhead	7.87+0.39	200+10	KC+ER
Auxiliary Steering Gear Bulkhead	2.76-3.94	70-100	---
Secondary Turrets)			
Face Plate	11.02	280	KC
Fore Side Plate	5.12	130	KC
Rear Side Plate	3.15	80	KC
Rear	3.15	80	KC
Fore Roof Plate	5.91	150	KC
Rear Roof Plate	4.13	105	KC
Upper Barbette	3.94-5.91	100-150	KC
Lower Barbette	1.57-2.76	40-70	---
Long Range AA Guns			
Turrets	0.47-1.57	12-40	---

Barbettes	1.57	40	---
Bow Section)			
Second Deck	0.47	12	ER
Armour Deck	2.36+0.39	60+10	OD+ER
Belt	2.76-5.12	70-130	KC
Transverse Bulkhead	2.36	60	---
Uptakes)			
Cylindrical Bulkheads	8.86	225	KC

Table 3.2

Richelieu – Armour Protection			
Designation	Thickness		Material
	in	mm	
Citadel)			
Main Belt (15.4°)	12.87+0.71	327+18	C
Lower Belt (21°)	12.87-6.97+0.71	327-177+18	C
Main Deck – Machinery/Center	5.91	150	NiCr
Main Deck – Machinery/Sides (Richelieu)	5.91	150	NiCr
Main Deck – Machinery/Sides (Jean Bart)	5.51	140	NiCr
Main Deck – Forward Magazines	6.69	170	NiCr
Second Deck – Center	1.57	40	NiCr
Second Deck – Sides (56°)	1.97	50	NiCr
Forward Transverse Bulkhead	6.50-9.17-13.98+0.71	165-233-355+18	C
Aft Transverse Bulkhead	6.50-9.17+0.71	165-233+18	C
Longitudinal Splinter Bulkhead	0.79	20	---
Torpedo Bulkhead – Machinery	1.18	30	S
Torpedo Bulkhead – Magazines	1.57-1.97	40-50	S
Magazine Bottom Deck	1.18	30	NiCr
Transverse Splinter Bulkheads	0.71	18	---
Main Turrets)			
Face Plate (20°)	16.93	430	C
Sides	11.81	300	C
Rear – Turret I.	10.63	270	C
Rear – Turret II.	10.24	260	C
Fore Roof Plate (9°)	7.68	195	C
Rear Roof Plate	6.69	170	C
Internal Longitudinal Bulkhead	1.77	45	---
Floor	2.17	55	---
Glacis	2.17+5.91	55+150	HTS
Overhang	2.17+1.97	55+50	HTS
Rangefinder Hoods	6.69	170	NiCr
Sight Hoods	4.53	115	NiCr
Barbettes)			
Upper	15.94+0.79	405+20	C
Lower	3.35	85	HTS
Conning Tower)			
Beam	13.39+0.79	340+20	C
Forward	13.39+0.79	340+20	C
Aft	11.02+0.79	280+20	C
Armoured Doors	11.02	280	S
Roof	6.69+0.79	170+20	C
Deck	3.94	100	S
Tube	6.30	160	S
Steering Gear Room, Shaft Tunnels)			
Turtle Deck Abaft Citadel	3.94	100	NiCr
Turtle Deck Abreast Steering Gear	5.91	150	NiCr
Aftermost Bulkhead	5.91	150	---
Forward Steering Gear Bulkhead	1.97	50	---
Secondary Turrets)			
Face Plate (25°)	5.12	130	---
Sides	2.76	70	---
Rear	2.36	60	---
Roof	2.76	70	---
Barbettes	3.94	100	---
Floor	1.18	30	---

Glacis	1.18+2.56	30+65	---
Sights	4.53	115	---
Rangefinder Hood	2.36-2.76	60-70	---
Hoistrings	5.51	140	---
Superstructure)			
Directors	0.79	20	---
Fire-Control Tower (Richelieu)	0.39	10	---
Fire-Control Tower (Jean Bart)	0.39-2.76	10-70	---
Uptakes)			
Bulkheads	0.79	20	---
Grating	11.81	300	---
Bow Section)			
Deck	1.57	40	NiCr

Table 3.3

Dimensions (Littorio)			
Designation	ft	m	% of Waterline Length
Citadel – Length	393.7	120	51.6
Magazine Deck – Length	239.5	73	31.4
Machinery Deck – Length	154.2	47	20.2
Lower Deck Aft – Length	160.8	49	21.1
Armour Deck/Belt Forward – Length	82.0	25	10.8
Main Armour Belt – Depth	14.4	4.4	---
Absorbing Cylinder – Diameter (Midship)	12.5	3.8	---
Torpedo Defense System – Depth (Midship)	23.7	7.22	---

Table 3.4

Dimensions (Richelieu)			
Designation	ft	m	% of Waterline Length
Citadel – Length	431.3	131.45	54.3
5.91" (150 mm) Main Deck – Length	268.0	81.70	33.8
6.69" (170 mm) Main Deck – Length	163.2	49.75	20.6
3.94" (100 mm) Lower Deck Aft – Length	106.6	32.50	13.4
5.91" (150 mm) Lower Deck Aft – Length	36.1	11.00	4.5
1.57" (40 mm) Lower Deck Forward – Length	165.7	50.05	20.9
Armour Belt – Depth Above Waterline	11.1	3.38	---
Armour Belt – Depth Below Waterline	8.5	2.58	---
Armour Belt – Total Depth	18.5	5.64	---
Side Protection – Max. Depth (Richelieu)	23.0	7.00	---
Side Protection – Max. Depth (Jean Bart)	27.1	8.27	---

Table 3.5

Immunity of Littorio Vs. Richelieu's Guns at 90 deg. (I.V.: 2,723 fps/830 mps) Based on Bu. Ord. Sk. 78841 – yds (m)			
Designation	Inner	Outer	Outer-Inner
Machinery	22,000 (20,117)	29,600 (27,066)	7,600 (6,949)
Magazines	22,000 (20,117)	31,900 (29,169)	9,900 (9,053)
Turrets	34,800 (31,821)	34,400 (31,455)	-400 (-366)
Barbettes	32,000 (29,260)	34,400 (31,455)	2,400 (2,194)
CT	45,604 (41,700)	24,500 (22,403)	-21,104 (-19,297)
Steering Gear	45,604 (41,700)	26,100 (23,866)	-19,504 (-17,834)

Table 3.6

Immunity of Littorio Vs. Richelieu's Guns at 90 deg. (I.V.: 2,575 fps/785 mps) Based on Bu. Ord. Sk. 78841 – yds (m)			
Designation	Inner	Outer	Outer-Inner
Machinery	19,000 (17,374)	28,300 (25,878)	9,300 (8,504)
Magazines	19,000 (17,374)	30,900 (28,255)	11,900 (10,881)
Turrets	41,065 (37,550)	33,900 (30,998)	-7,165 (-6,552)
Barbettes	28,000 (25,603)	33,900 (30,998)	5,900 (5,395)
CT	41,065 (37,550)	21,600 (19,751)	-19,465 (-17,799)
Steering Gear	41,065 (37,550)	24,100 (22,037)	-16,965 (-15,513)

Table 3.7

Immunity of Richelieu Vs. Littorio's Guns at 90 deg. (I.V.: 2,789 fps/850 mps) Based on Bu. Ord. Sk. 78841 – yds (m)			
Designation	Inner	Outer	Outer-Inner
Machinery	25,300 (23,134)	34,900 (31,913)	9,600 (8,878)
Magazines	25,300 (23,134)	36,900 (33,741)	11,600 (10,607)
Turrets	32,800 (29,992)	32,500 (29,718)	-300 (-274)
Barbettes	28,100 (25,695)	32,500 (29,718)	4,400 (4,023)
CT	33,900 (30,998)	35,300 (32,278)	1,400 (1,280)
Steering Gear	46,807 (42,800)	33,200 (30,358)	-13,607 (-12,442)

Figure 3.1

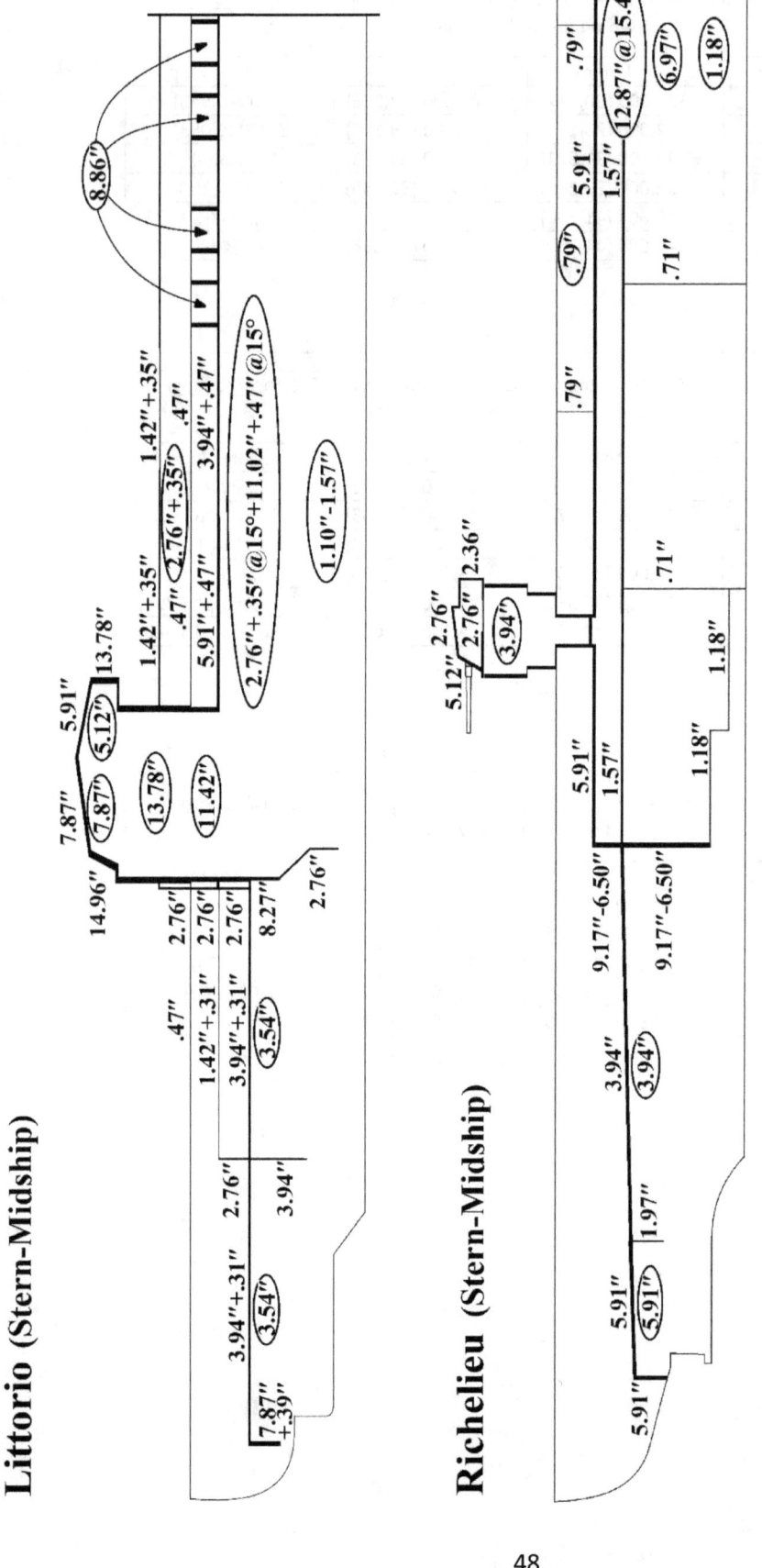

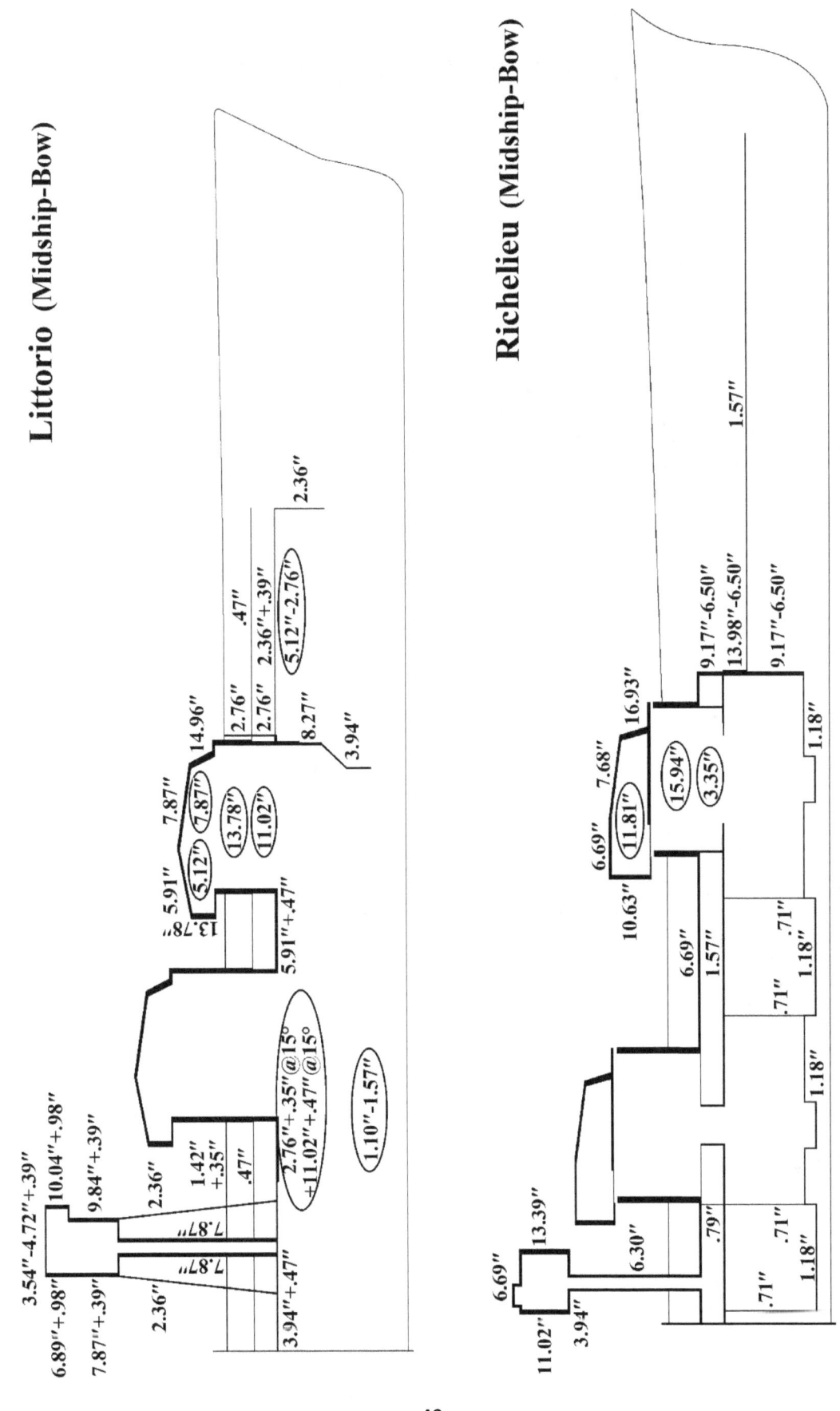

Figure 3.2

Steering Gear/Shafts/Stern

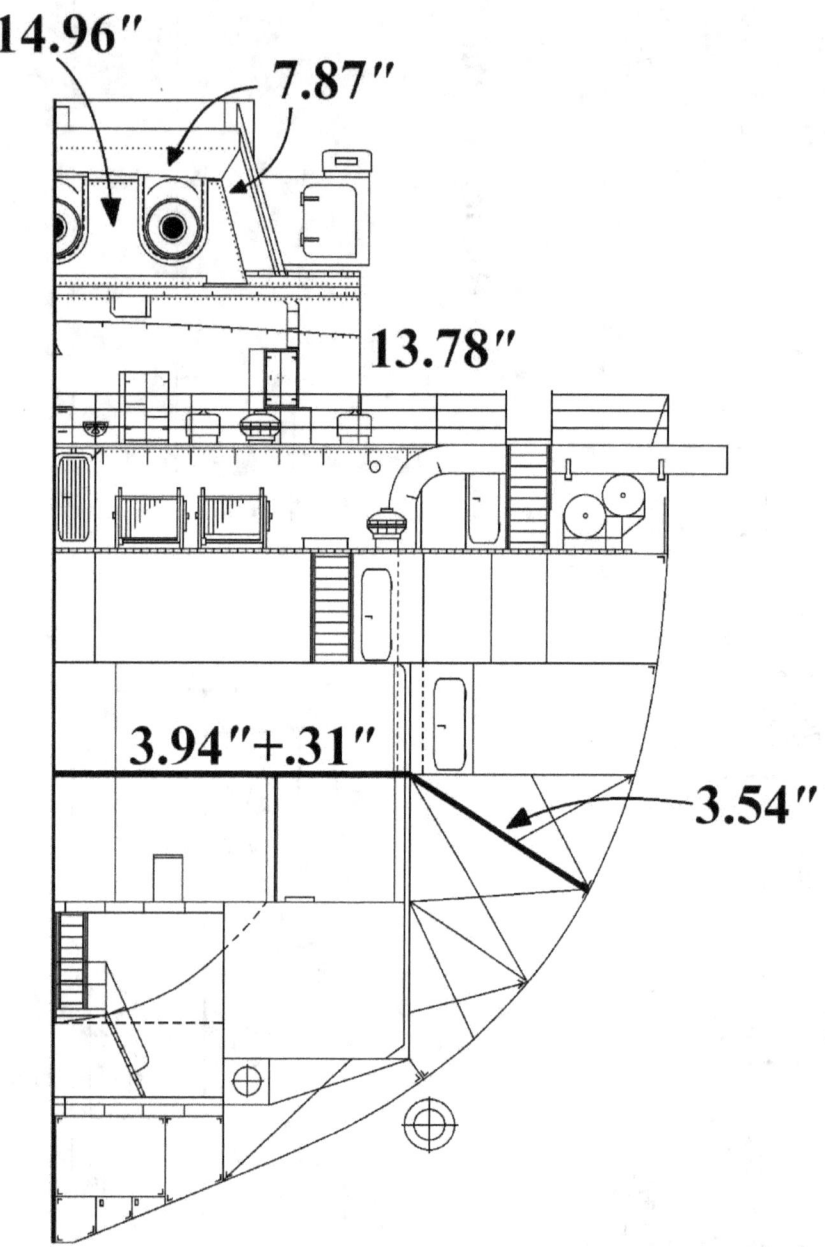

Littorio

Steering Gear/Shafts/Stern

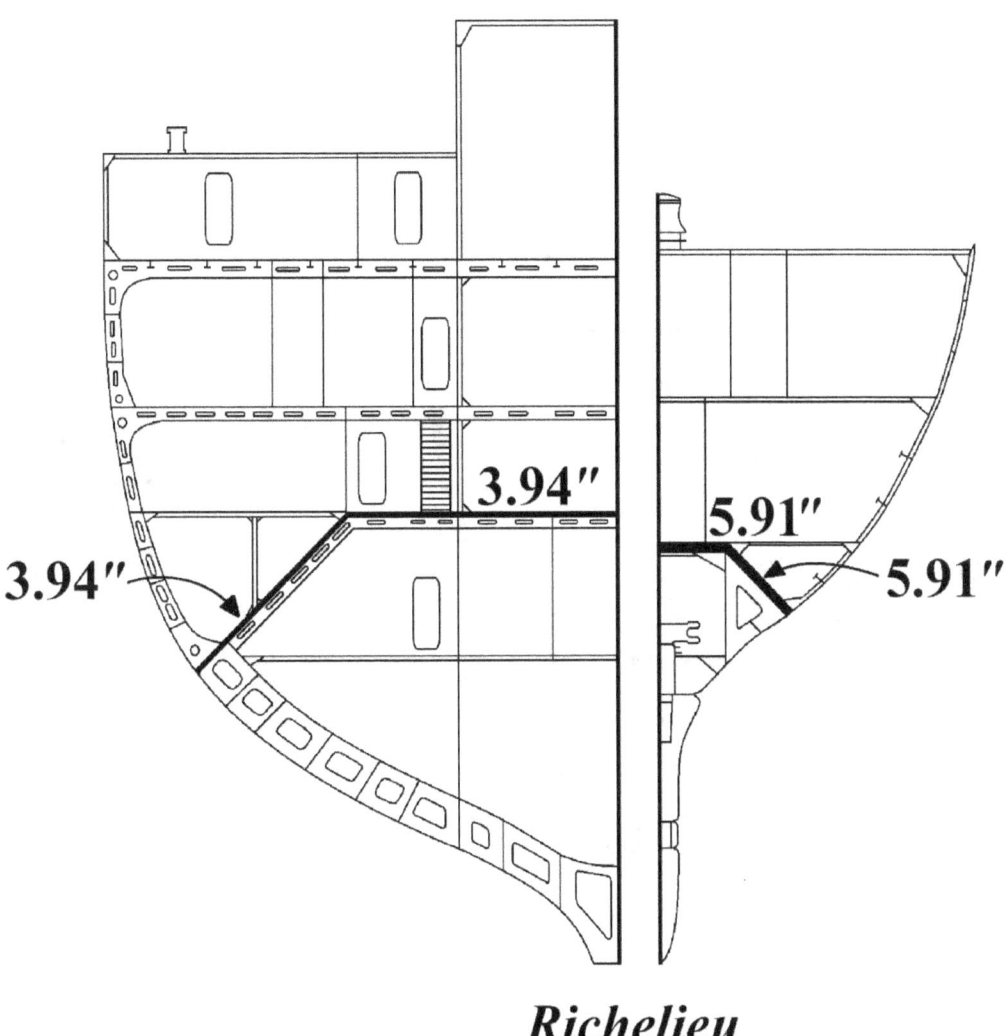

Richelieu

Machinery

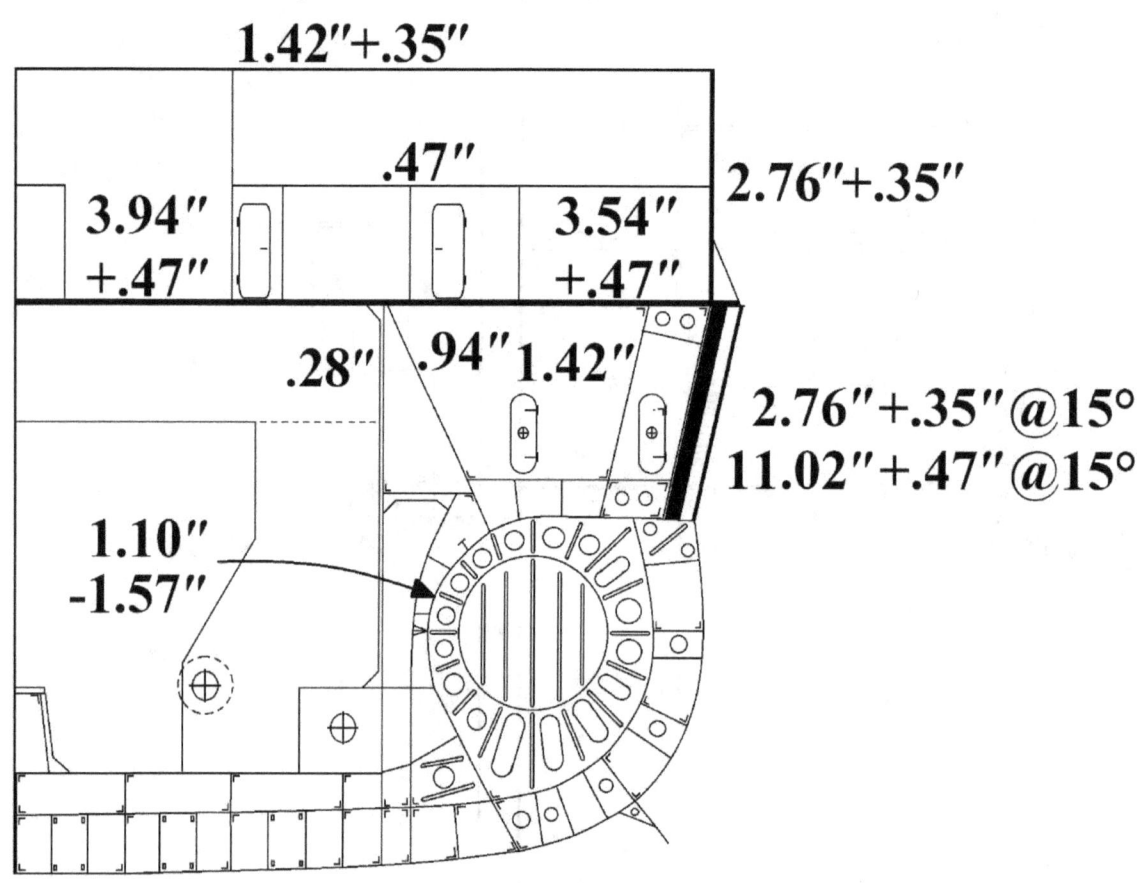

Littorio

Machinery

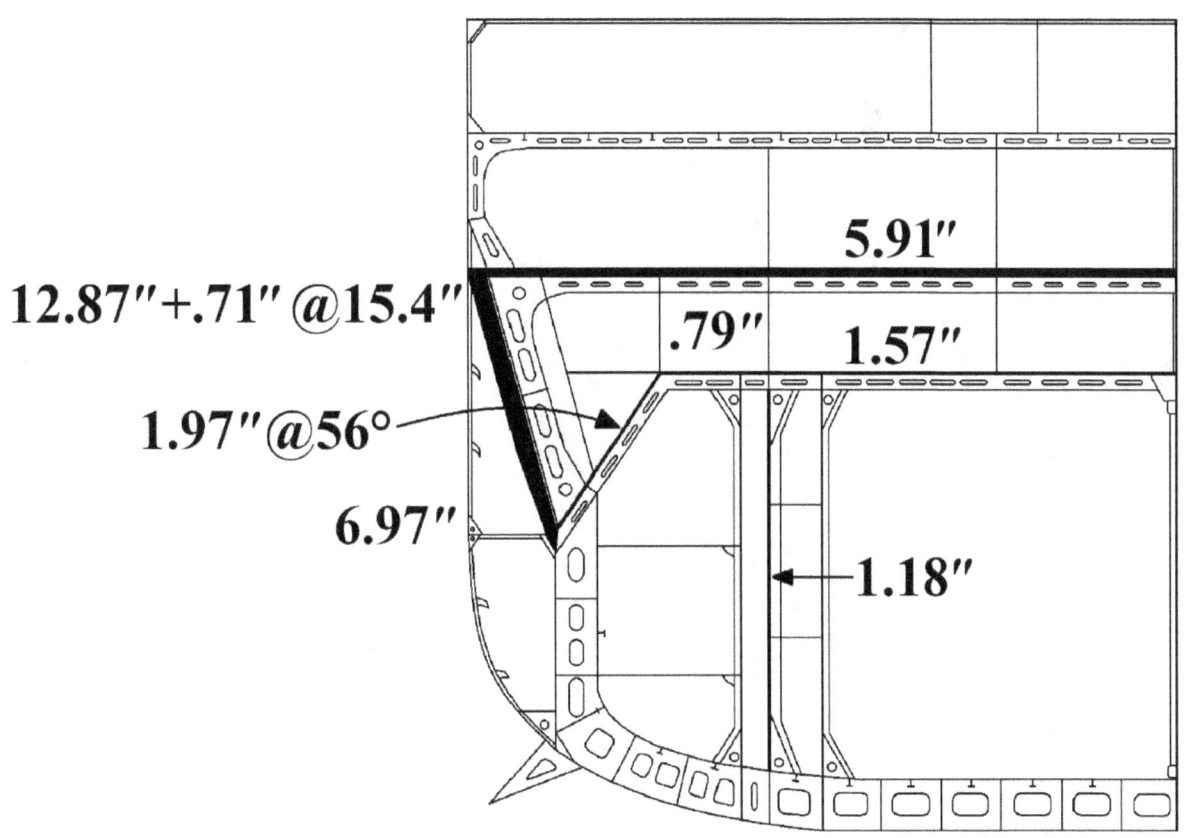

Richelieu

Magazines

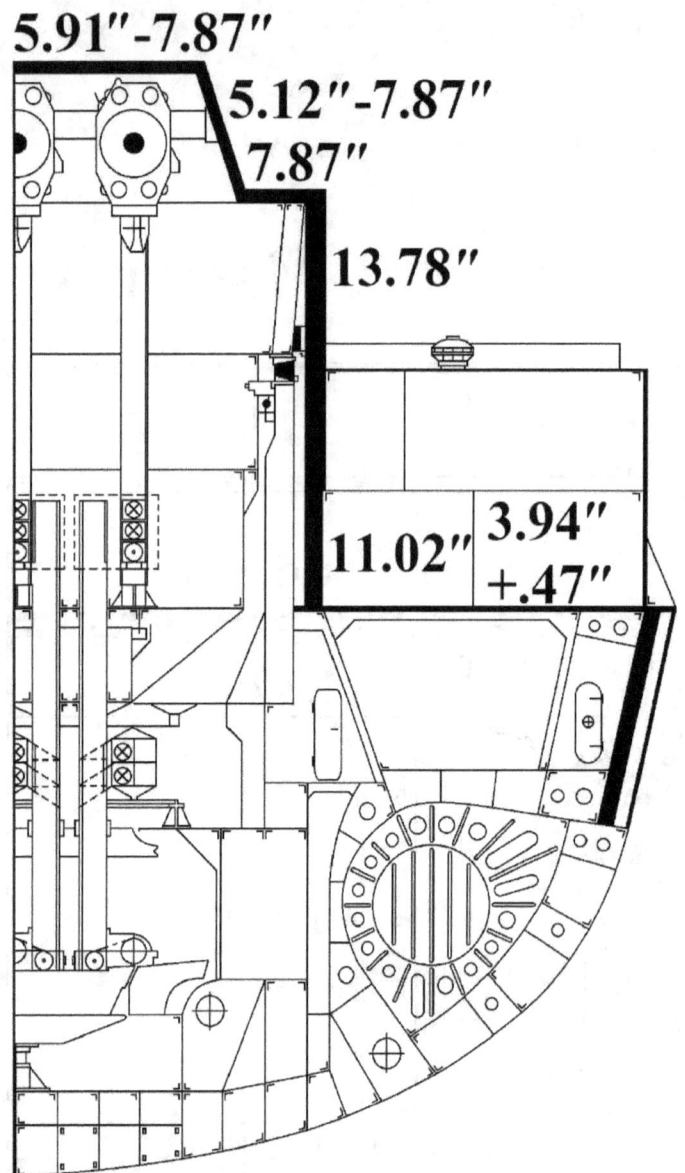

5.91"-7.87"
5.12"-7.87"
7.87"
13.78"
11.02" 3.94"
        +.47"

Littorio

Magazines

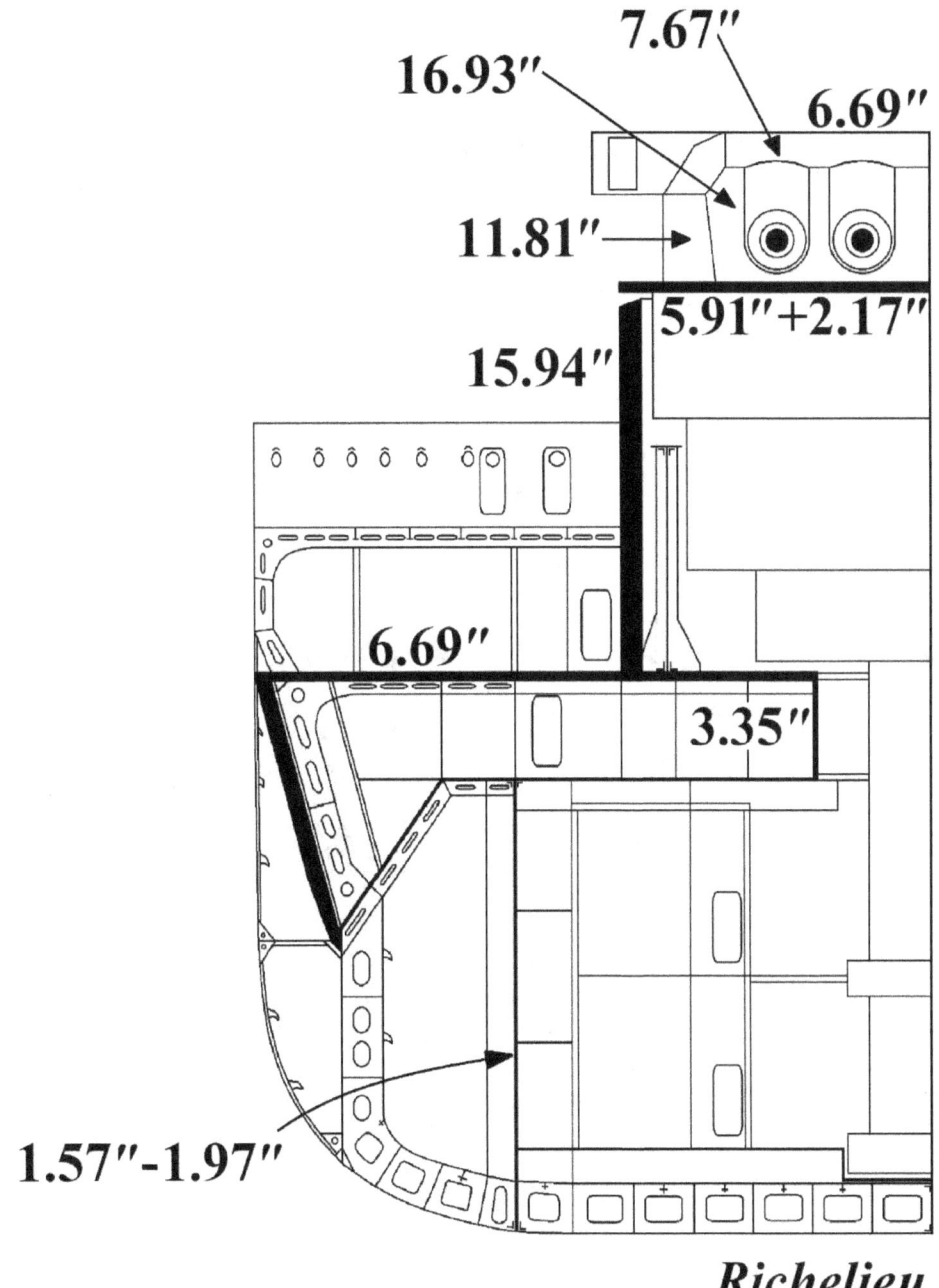

Richelieu

Secondary Armament

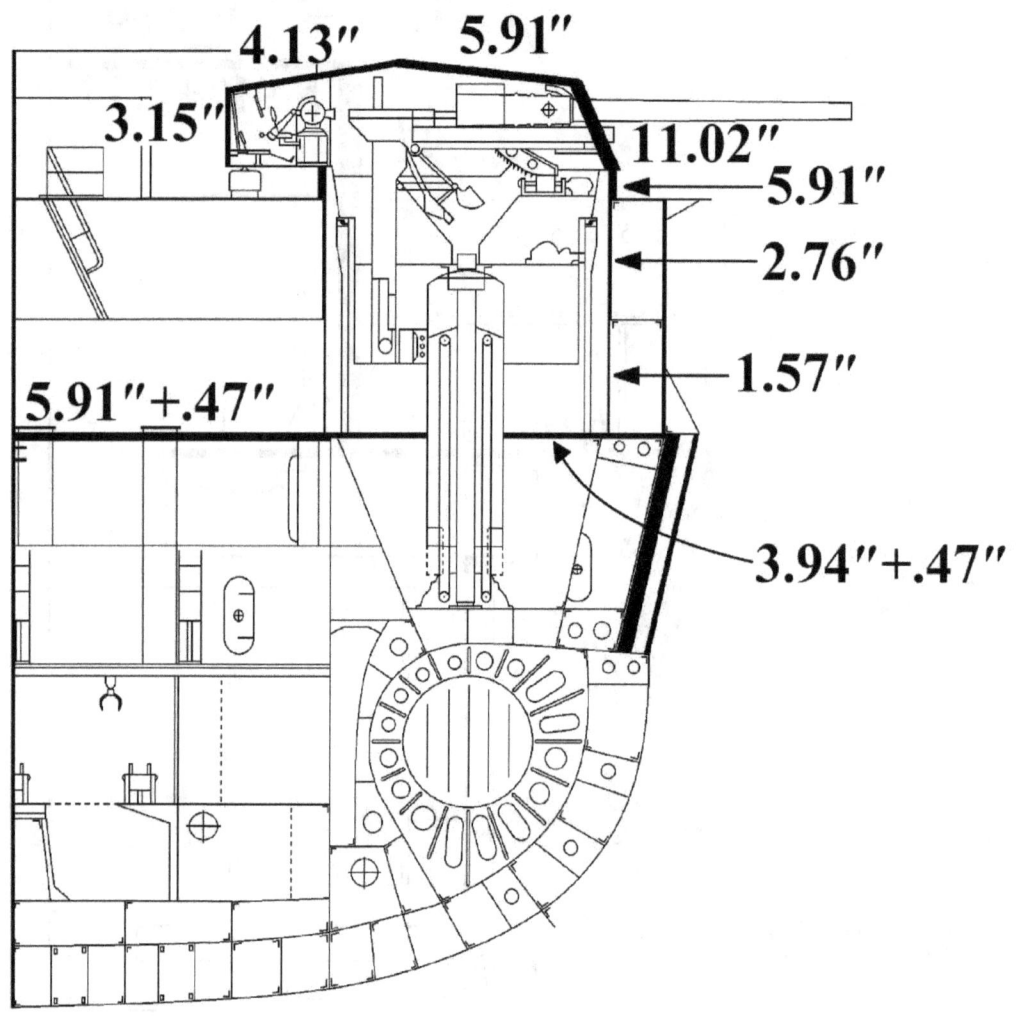

Littorio

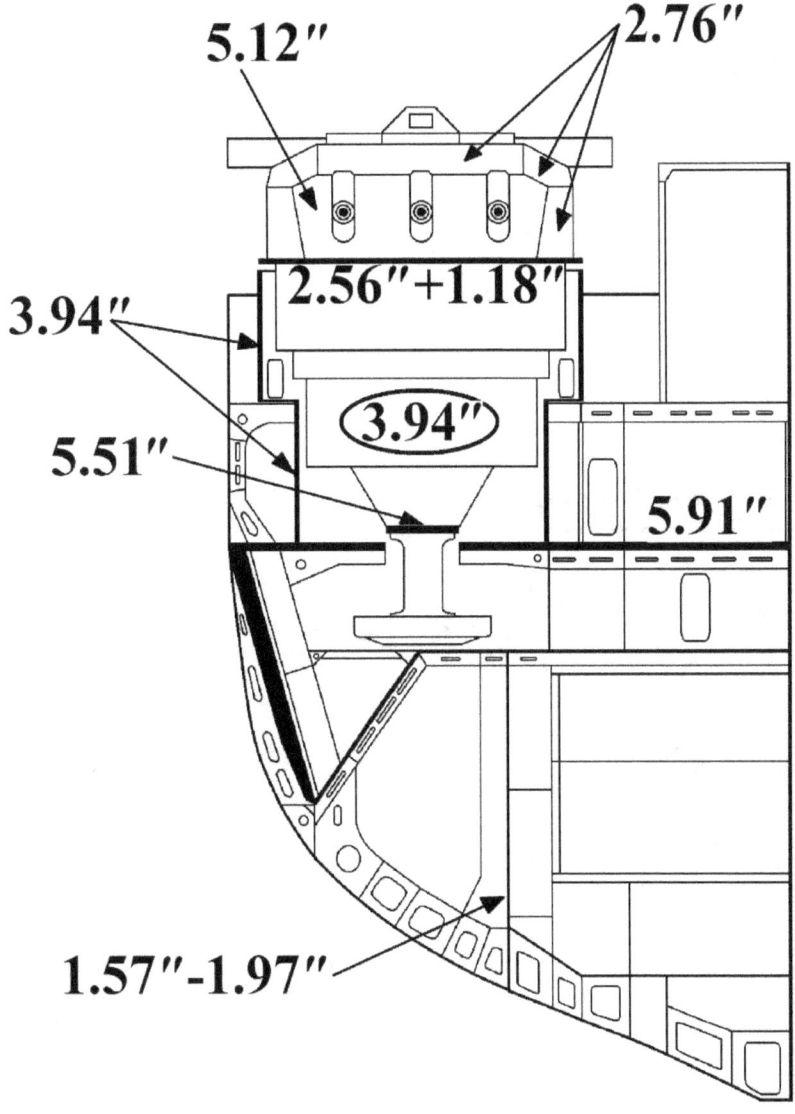

Richelieu

Bow

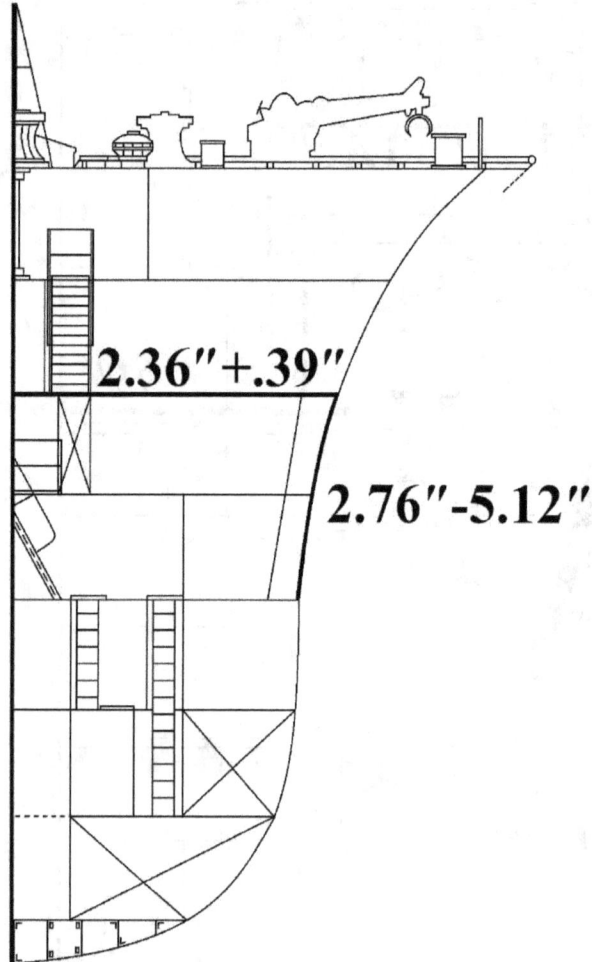

Littorio

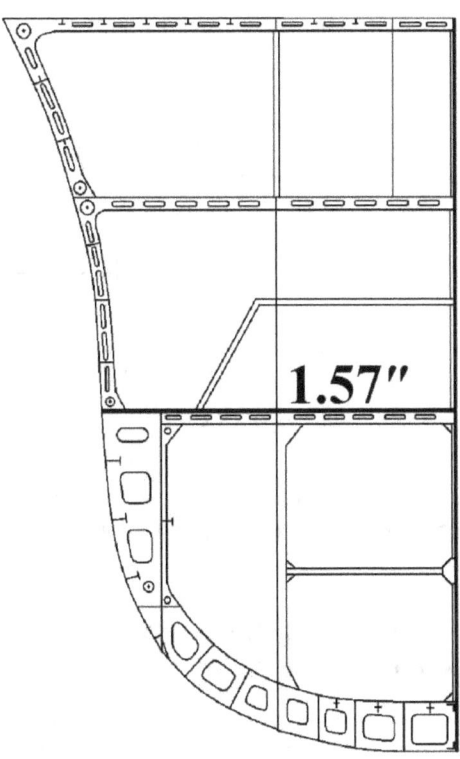

Figure 3.3

Immunity Zones at 90 deg. Target Angle
(Based on Bu. Ord. Sk. 78841)

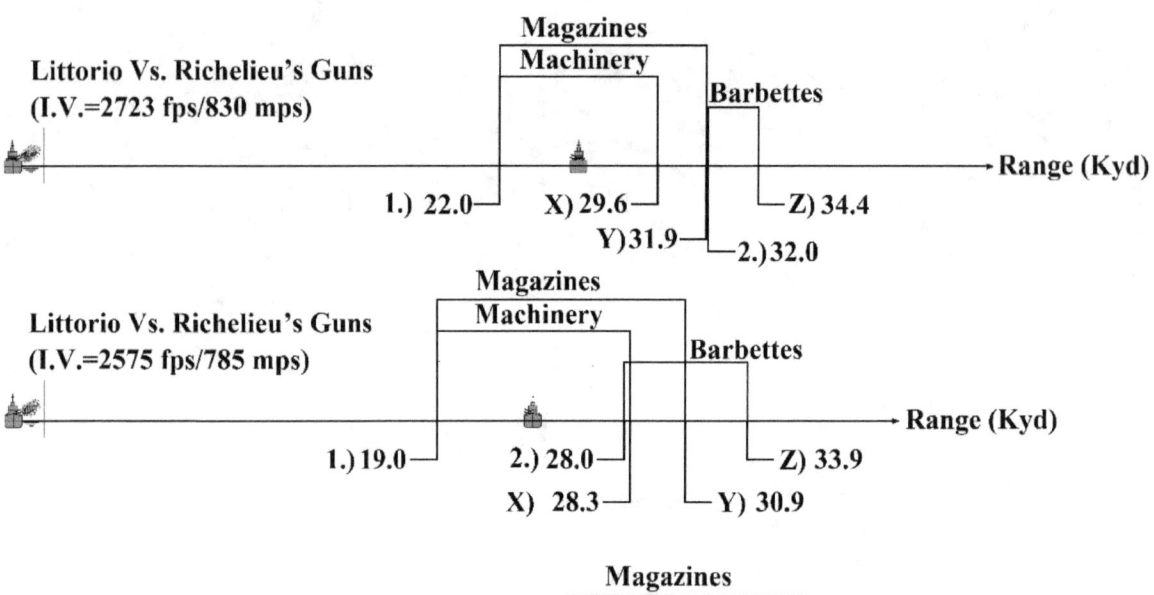

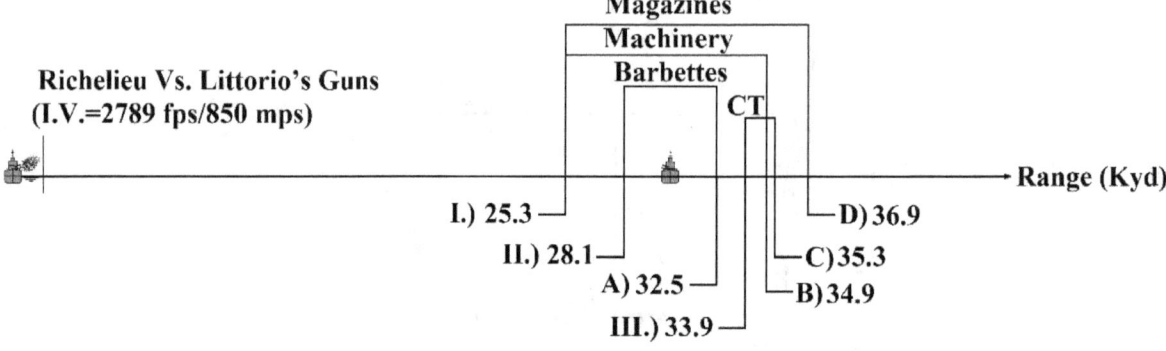

1.) 2.76"+.35"@15° Belt+11.02"+.47"@15° Belt+1.42"@15° Bhd.+.94"@26° Bhd.+.28" Bhd.
2.) 13.78" Barbette (Beam)

X) 2.76"+.35" Belt+3.54"+.47" Deck+.94"@26° Bhd.+.28" Bhd.
Y) 2.76"+.35" Belt+3.94"+.47" Deck+.94"@26° Bhd.+.28" Bhd.
Z) 7.87"@7° Turret (Fore Roof Plate)

I.) 12.87"@15.4° Belt+1.97"@56° Slope + 1.18" T. Bhd.	A) 7.69"@9° Turret (Fore Roof Plate)
II.) 15.94" Barbette (Beam)	B) 5.91" Deck+1.57" Deck
III.) 13.39" CT (Beam)	C) 6.69" CT (Roof)
	D) 6.69" Deck+1.57" Deck

Figure 3.4

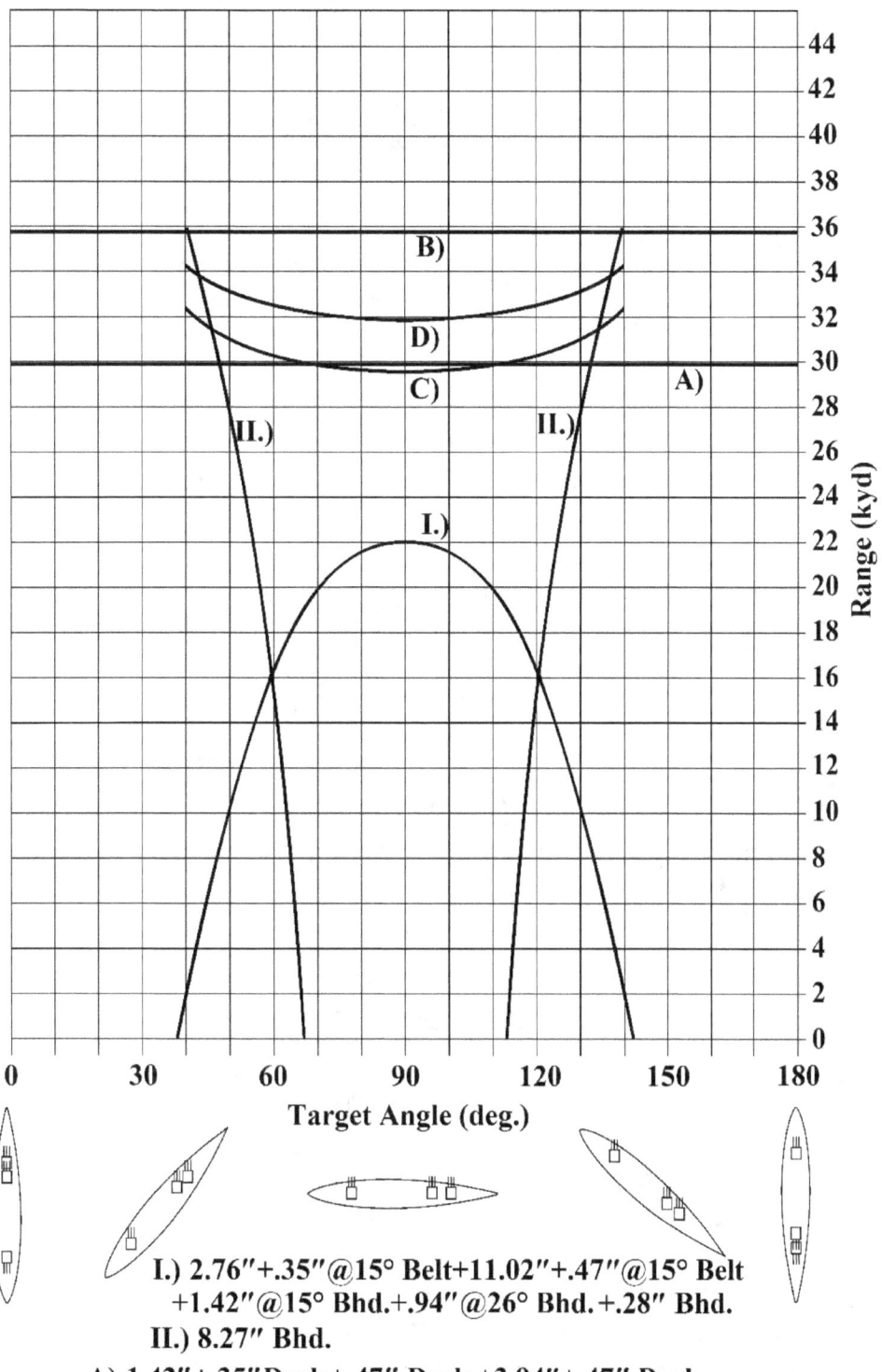

Immunity of Littorio's Citadel Vs. Richelieu's Guns
(I.V.=2723 fps/830 mps, Based on Bu. Ord. Sk. 78841)

I.) 2.76"+.35"@15° Belt+11.02"+.47"@15° Belt +1.42"@15° Bhd.+.94"@26° Bhd.+.28" Bhd.
II.) 8.27" Bhd.
A) 1.42"+.35"Deck+.47" Deck +3.94"+.47" Deck
B) 1.42"+.35"Deck+.47" Deck +5.91"+.47" Deck
C) 2.76"+.35"Belt+3.54"+.47" Deck+.94"@26° Bhd.+.28" Bhd.
D) 2.76"+.35"Belt+3.94"+.47" Deck+.94"@26° Bhd.+.28" Bhd.

Figure 3.5

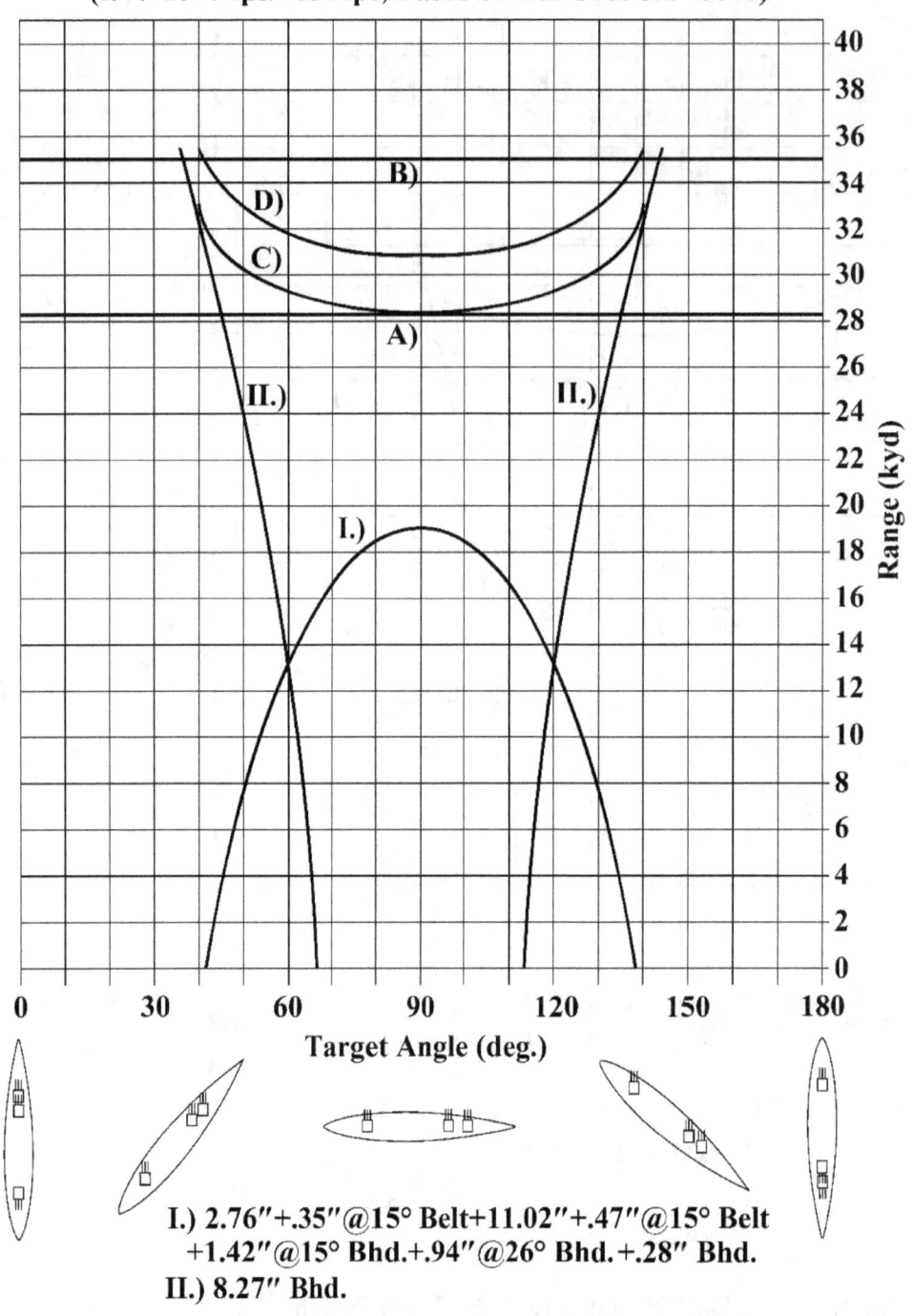

Immunity of Littorio's Citadel Vs. Richelieu's Guns
(I.V.=2575 fps/785 mps, Based on Bu. Ord. Sk. 78841)

I.) 2.76"+.35"@15° Belt+11.02"+.47"@15° Belt
+1.42"@15° Bhd.+.94"@26° Bhd.+.28" Bhd.
II.) 8.27" Bhd.

A) 1.42"+.35"Deck+.47" Deck +3.94"+.47" Deck
B) 1.42"+.35"Deck+.47" Deck +5.91"+.47" Deck
C) 2.76"+.35"Belt+3.54"+.47" Deck+.94"@26° Bhd.+.28" Bhd.
D) 2.76"+.35"Belt+3.94"+.47" Deck+.94"@26° Bhd.+.28" Bhd.

Figure 3.6

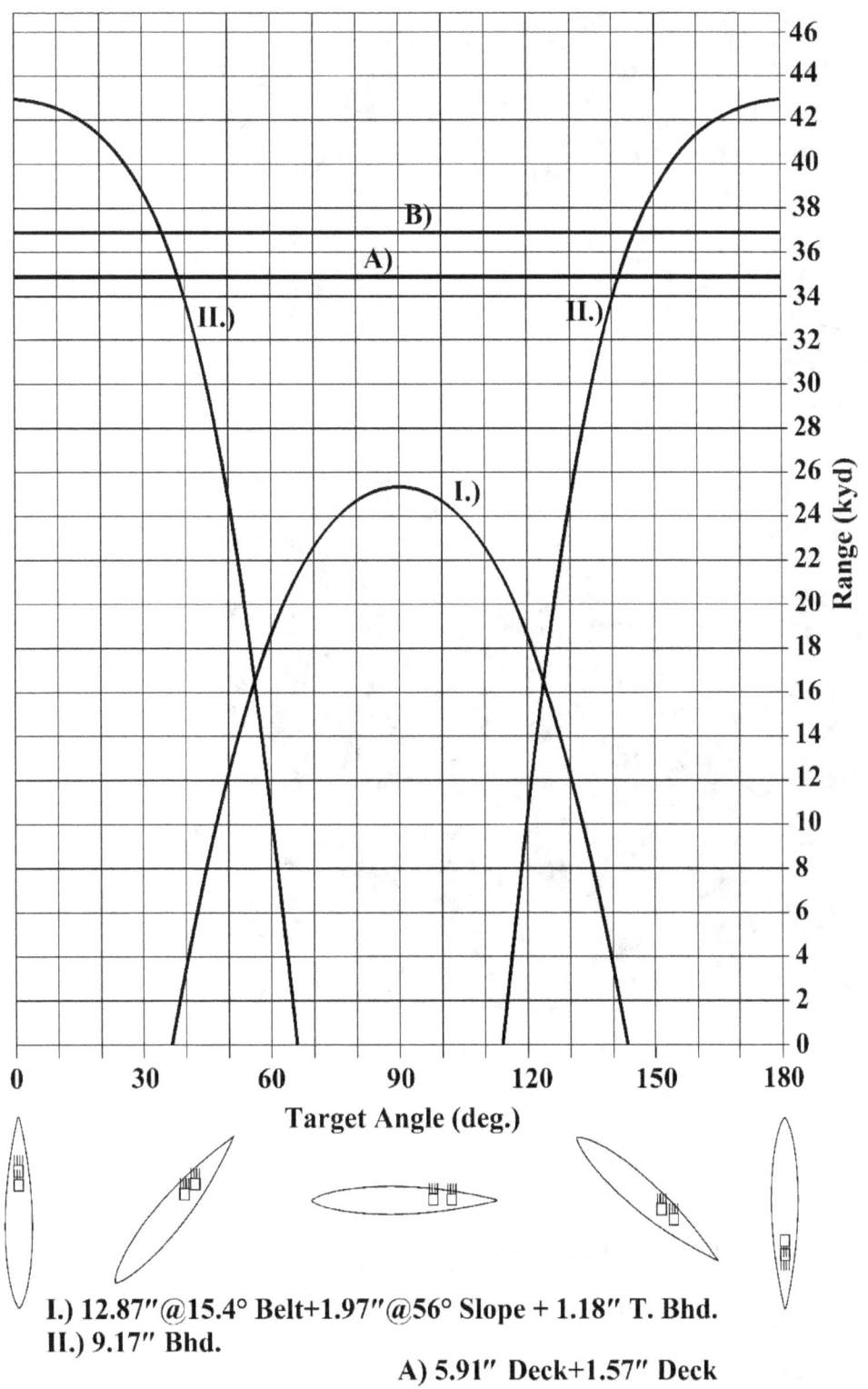

I.) 12.87"@15.4° Belt+1.97"@56° Slope + 1.18" T. Bhd.
II.) 9.17" Bhd.

A) 5.91" Deck+1.57" Deck
B) 6.69" Deck+1.57" Deck

Figure 3.7

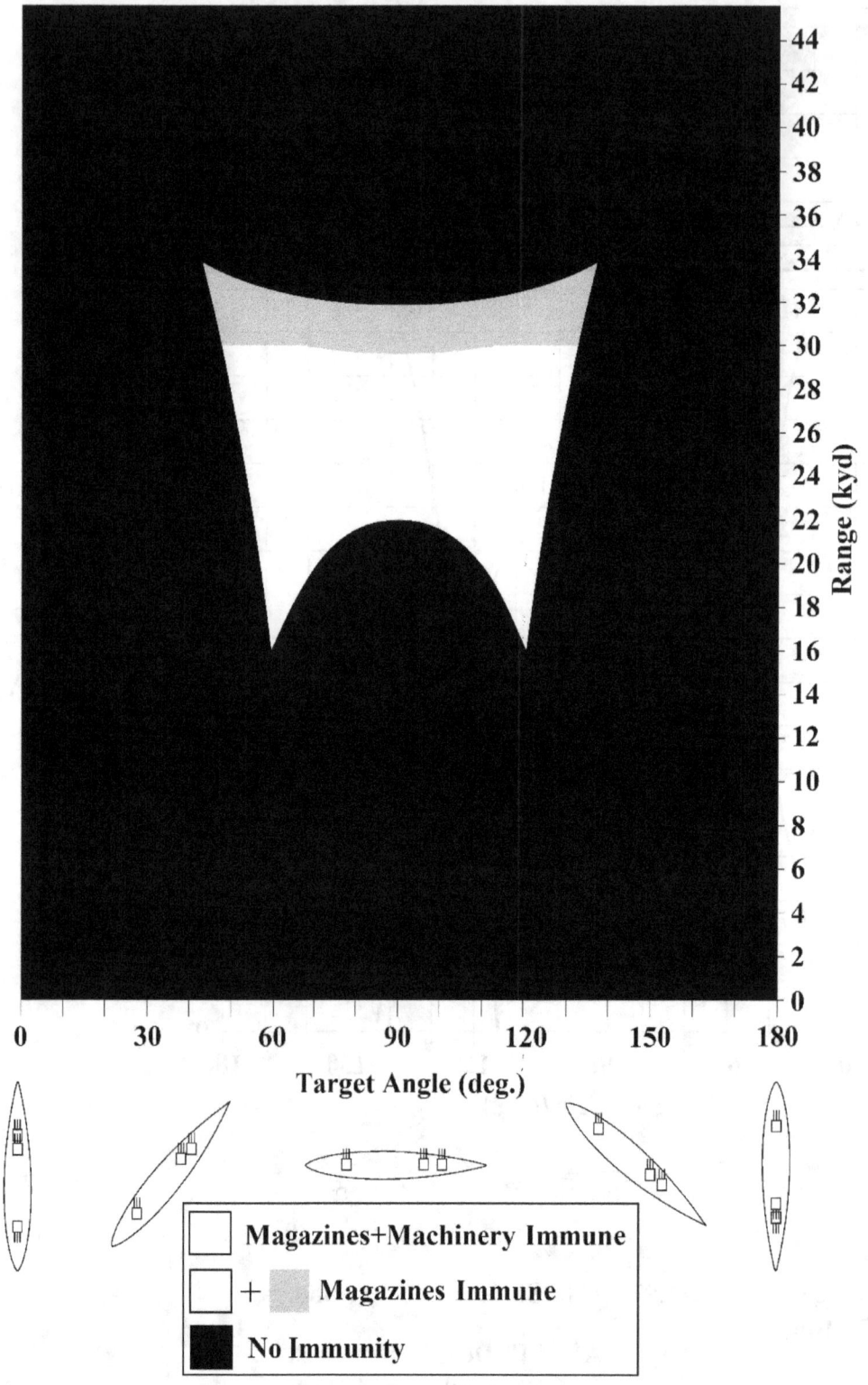

Figure 3.8

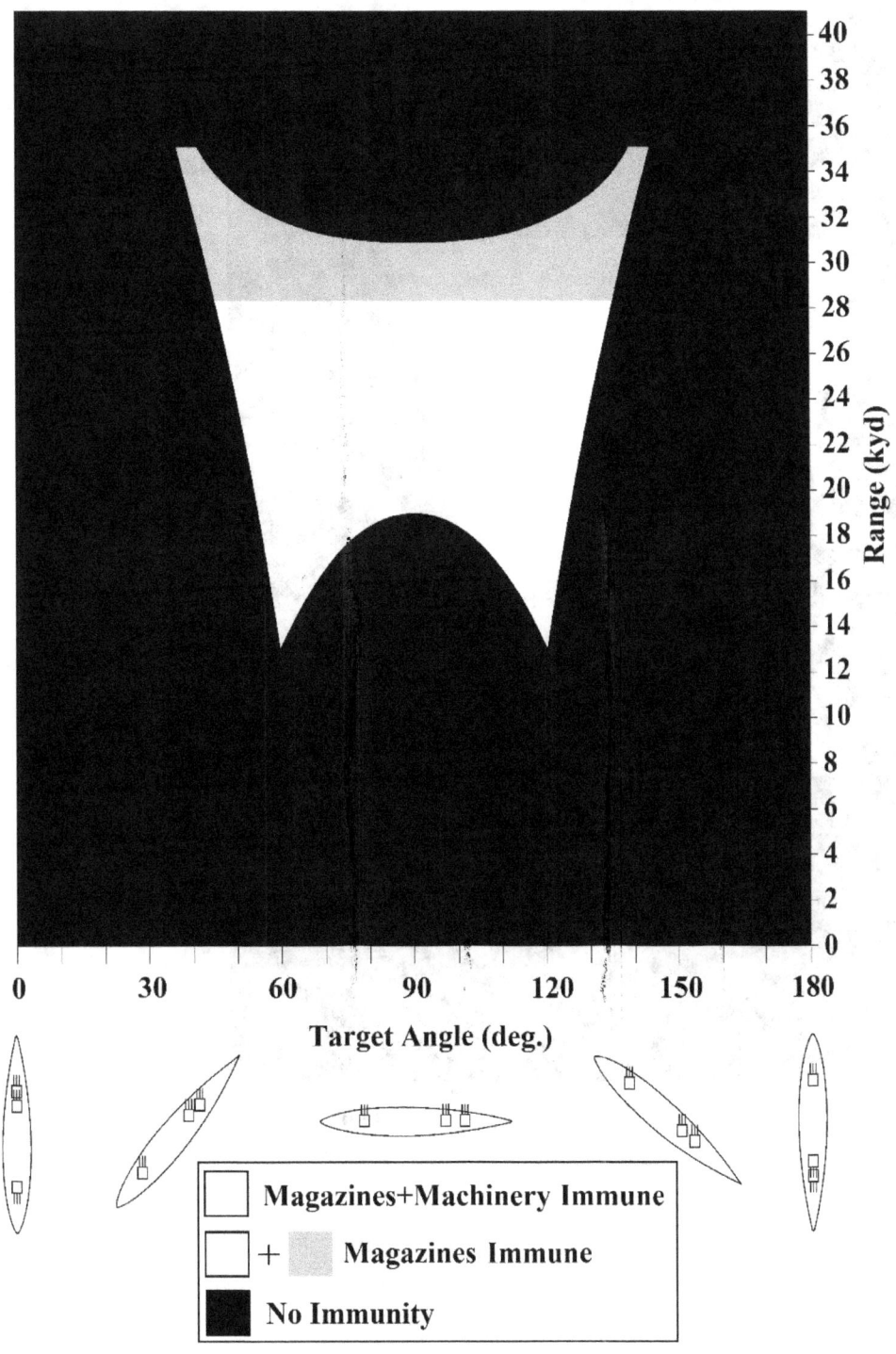

Figure 3.9

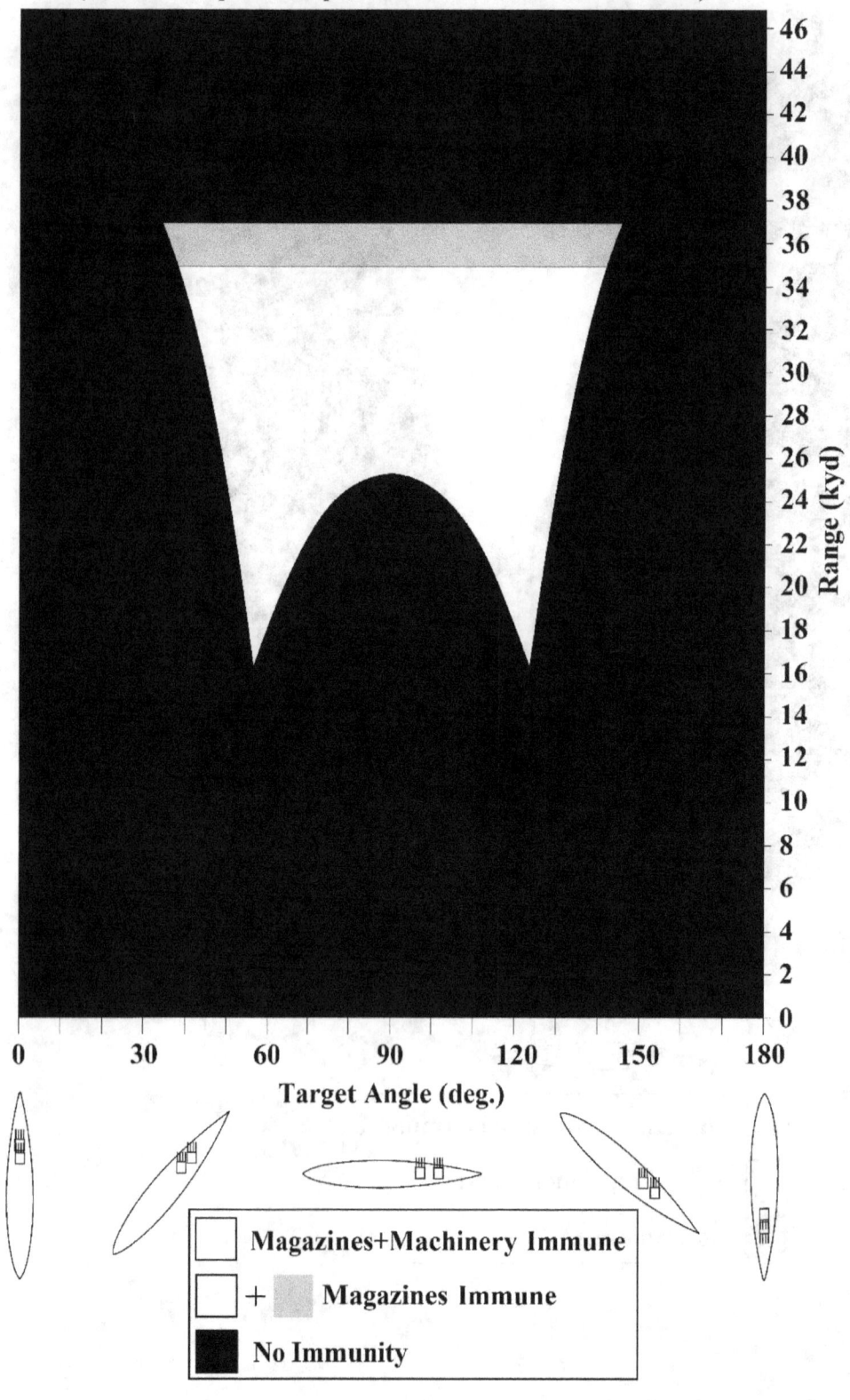

Figure 3.10

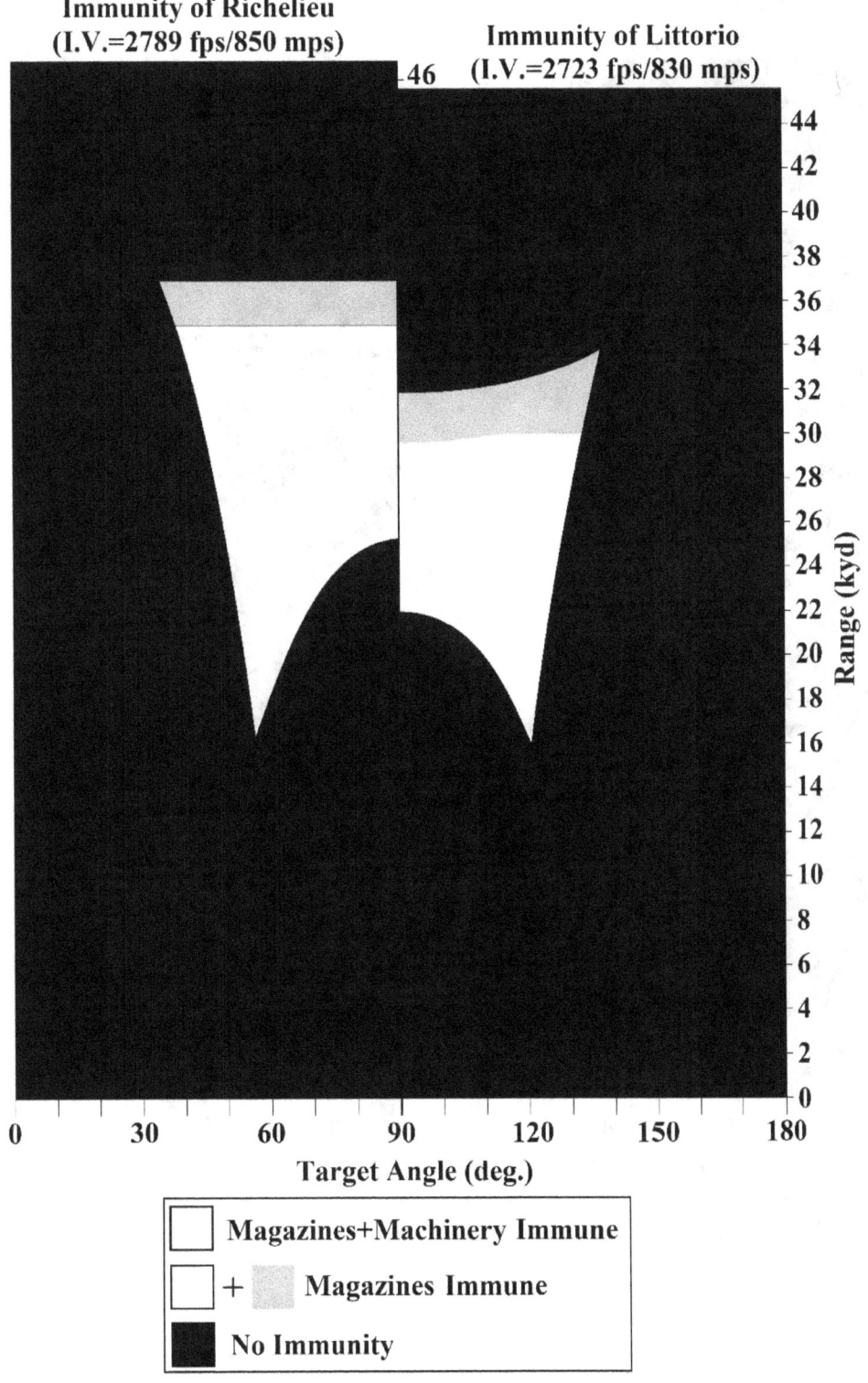

Figure 3.11

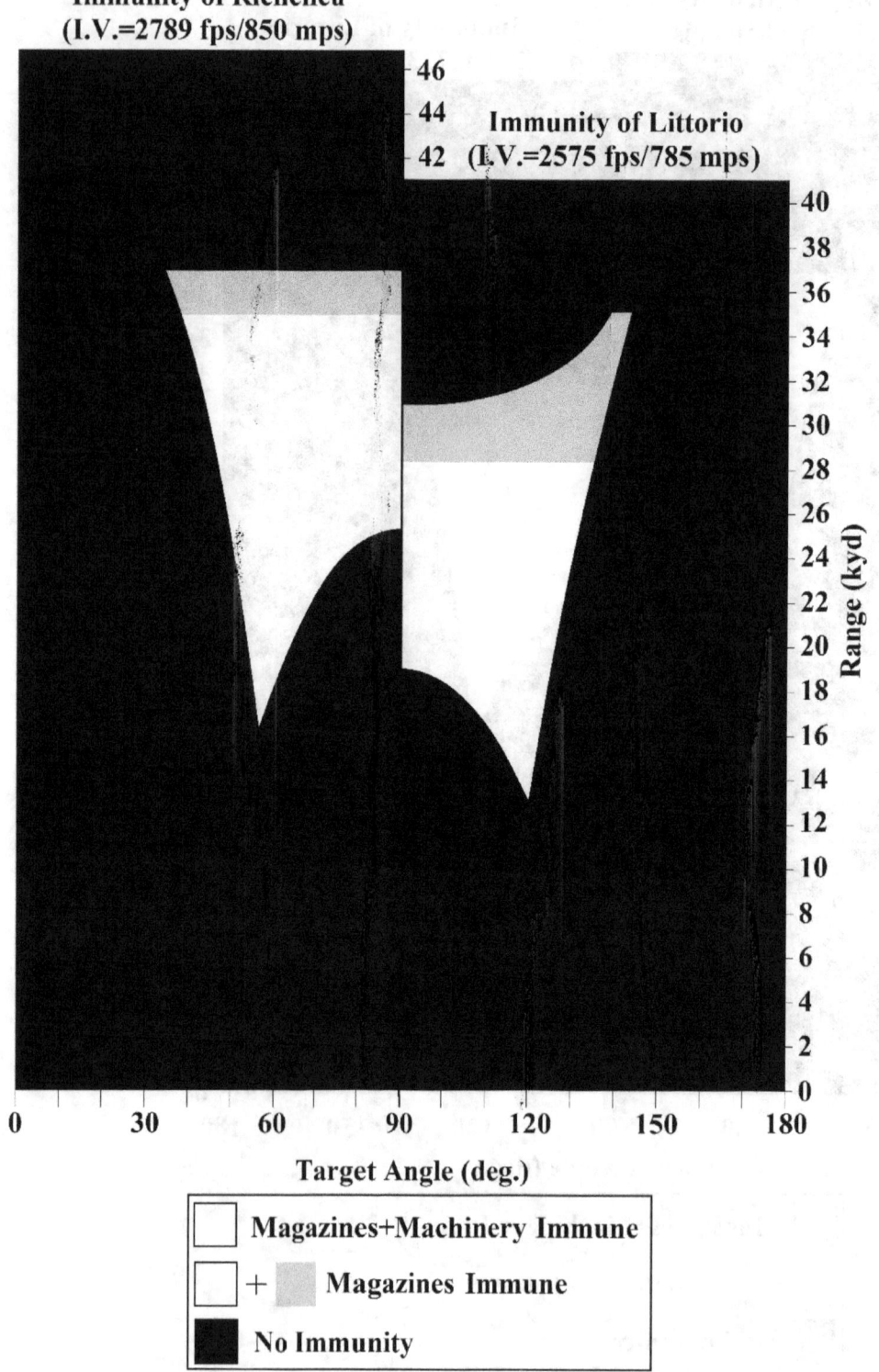

Figure 3.12

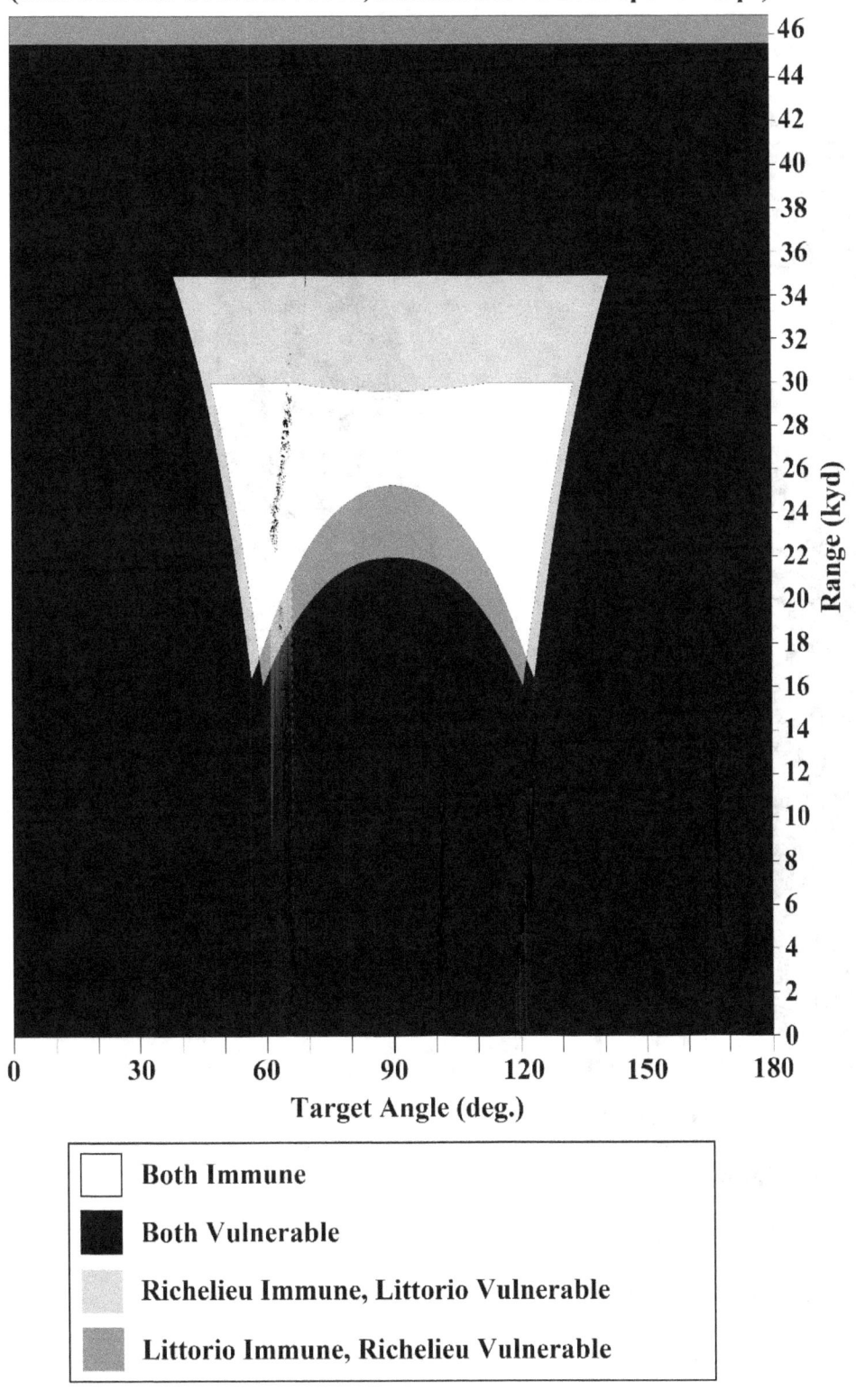

Figure 3.13

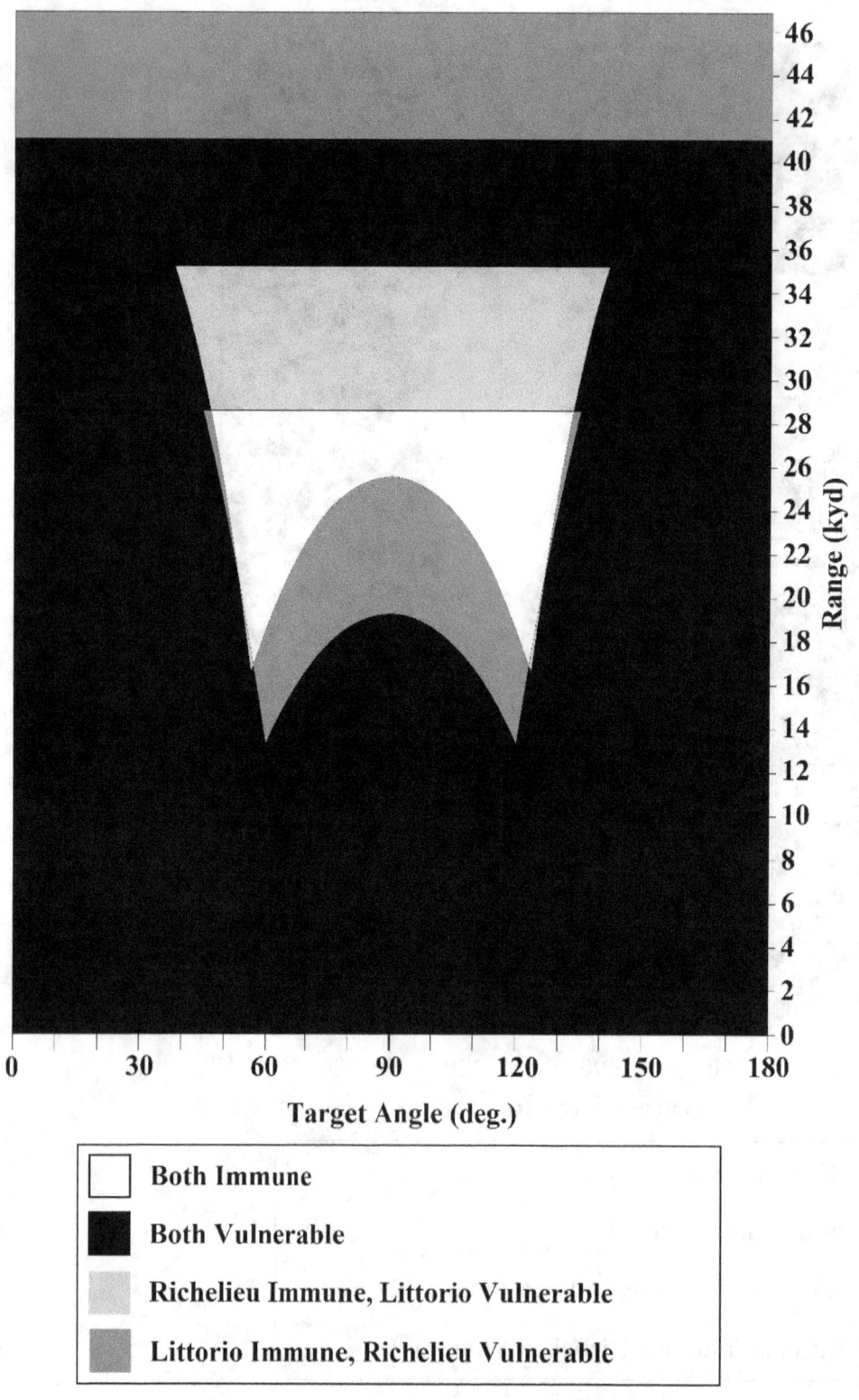

Conclusion

In light of the fact that the ballistics, ammunition, replenishment system and fire-control sensors of Richelieu's main guns were altered during the early 1940s, we separate the following discussion into two parts. The first part deals with Littorio facing Richelieu as designed, while in the second part we shall assume that Richelieu has been refitted.

I.) <u>Littorio VS Richelieu (as designed)</u>

Though it would seem that Richelieu would have dominated at 30-35 kyd (27.4-32.0 km) on account of her thicker armour deck being immune to the Italian guns, her deeper side armour providing more reliable protection against diving shells and the director rangefinders on the French side outclassing those of the Italian ship, the probability that the engagement would have taken place at such ranges is low in light of actual, historical battleship engagements. Indeed, no battleship gun ever scored hits outside 27 kyd (24.7 km) against moving targets, and most engagements, including those fought between contemporary battleships, occurred at shorter ranges than this. Not to mention that Richelieu's foretop main battery rangefinder was elevated high enough to track a target as high as Littorio up to about only 30 kyd (27.4 km). Admittedly, the ship's spotting aircraft could provide fire-control adjustments at any range.

Another important question is; how accurate was the intel the French possessed on Littorio's technical characteristics? The Italians were not nearly as secretive as, for instance, the Japanese; Russian battleship plans of the period indicate that this nation had detailed information about the Pugliese-type torpedo defense system, while ADM reports reveal that by 1937 the Admiralty knew that the side armour of the Italian battleships was to be composed of a spaced assembly. It was also well-known that the Italians preferred high-velocity guns which are quite inefficient against an armour deck as thick as Richelieu's.

Therefore, it is not impossible that the French – assuming that they knew that the armour deck of the Italian battleship was considerably lighter than that of their ship – would have attempted to engage at extreme range, but providing accurate fire-control would have been complicated. It would have been at least equally difficult for Richelieu to keep her distance, considering that the Italians would have certainly attempted to engage at shorter ranges than this.

Gunnery trials of the Italian ships and the estimated immunity provided by their vertical armour indicate that this navy anticipated a decisive gun range of circa 22 kyd (20.1 km). Although Richelieu was marginally faster, sailing in the opposite direction would have been the only way to keep her distance if Littorio were to sail directly towards her, which means that she could not have returned Littorio's fire at all, whilst the Italian ship could have utilized two-thirds of her firepower; not to mention that this would have left Richelieu's aft transverse bulkhead, and consequently secondary gun magazines, vulnerable.

It should also be noted that the initial velocity of Richelieu's guns, the shape of her domestic AP projectiles and her sloped, cemented armour turret roof plates hint that engagements at such ranges were hardly what the designers of the French ship originally had in mind. Based on the designed immunity of Richelieu, it appears unlikely that the French expected an optimal battle range significantly greater than the Italians.

Finally, based on the results of the first hit scored by USS Massachusetts against Jean Bart, Richelieu's immunity outside 30 kyd (27.4 km) might not have been as reliable against Littorio as indicated by Bu. Ord. Sk. 78841.

In short, Richelieu's superiority outside 30 kyd (27.4 km) was shaky and impracticable.

Ranges corresponding to mutual immunity, that is 26-29 kyd (23.8-26.5 km) at 90 deg. target angle, would have also been more favorable to Richelieu, considering that diving projectiles would have been more dangerous for Littorio, owing to her shallower side armour and the enhanced underwater performance of the French shells. The fire-control efficiency of Richelieu would have probably been greater too, as a virtue of her superior director optics. Hits scored by her would have also been more destructive as the French shells were more bounteously filled with explosives.

At 22-25 kyd (20.1-22.9 km)/70-90 deg. Littorio might have had an edge. Not only did her vertical armour have a higher limit velocity, but it is also unlikely that, even if Richelieu's projectiles had enough energy to achieve penetration, they would have remained fit to burst inside Littorio's hull. Furthermore, at shorter ranges the visual horizon of turret rangefinders is sufficient to track battleship sized targets, which is more favorable to Littorio, taking into account the greater disparity between the size and number of her turret and director rangefinders. While Littorio definitely had ballistic superiority under these conditions, she was still more threatened by underwater hits than her foe. Furthermore, Littorio's ballistic superiority would have been relevant only if the target angle was no less than circa 70 deg.

At a 60 deg. target angle, the zone of mutual immunity extends to ranges as short as 18 kyd (16.5 km), and it is quite probable that the actual engagement would have taken place under such conditions, that is somewhere outside 18 kyd (16.5 km) with the ships inclined at circa 60 deg. with respect to the plane of fire. Both ships being immune and their fire-control sensors having arguably comparable efficiency, it would have been exceedingly difficult to reach a decisive outcome. Richelieu, however, would have still been able to deliver more explosives and Littorio would have still been more vulnerable against underwater hits.

Inside the zone of mutual vulnerability, that is inside 22 kyd (20.1 km)/90 deg or 16 kyd (14.6 km)/60 deg, Littorio would have probably had the upper hand. Although both ships were vulnerable in theory, Littorio's side armour might have broken up incoming shells.

At this range, harassing fire would have been possible too. The French medium caliber guns had an overall edge, but this was compensated by Littorio's more extensive medium armour plating.

Note that neither ship's armament, conning position or steering ability was adequately protected against the shellfire of the other.

As can be seen, this would have been a particularly tough battle for both sides, with neither ship having a truly decisive superiority, but long ranges and the zone corresponding to mutual immunity would have been more favorable to Richelieu, whilst at shorter ranges it seems more likely that Littorio would have had an edge.

II.)   <u>Littorio VS Richelieu (refitted)</u>

With their muzzle velocity decreased, the French guns could defeat Littorio's horizontal protection beyond circa 28 kyd (25.6 km). Unfortunately for the French, the British Type 284P gunnery radar and the U.S. SF type surface search radar had a maximum range of only 24 kyd (21.9 km). Hence, they still would have had to rely on director optics at very long ranges, albeit in 1945 Richelieu received an SG type surface search radar whose detection range was roughly the same as the max. range of the ship's main guns. However, this set was not specifically designed for fire-control and the ship no longer carried spotting aircraft.

On the other hand, Richelieu's accuracy was boosted by the introduction of firing delays, which nearly halved the dispersion of her main battery salvos. Also, as a result of the alteration of her replenishment system, her rate of fire was increased.

Littorio's vertical protection, however, was now immune against the slower U.S. projectiles outside approximately 19 kyd (17.4 km), whilst she could still pierce Richelieu's side armour up to about 25 kyd (22.9 km), assuming 90 deg. target angle in both instances.

As can be seen, both ships had more distinct advantages under conditions more favorable to them after Richelieu's refit. At 90 deg. target angle, Richelieu was now superior at 28-35 kyd (25.6-32.0 km), Littorio at 19-25 kyd (17.4-22.9 km); both immune at 25-28 kyd (22.9-25.6 km), but neither inside 19 kyd (17.4 km) and outside 35 kyd (32.0 km).

With accurate intel on her adversary and successful fire-control, Richelieu could have, in theory, defeated Littorio at very long ranges. However, the outer limit of the Italian ship's immunity was, in truth, not sufficiently diminished by the alteration of Richelieu's main armament to make this scenario likely. Littorio was still immune up to very long ranges, Richelieu's fire-control efficiency at this range was not materially improved by her refits, and the Italians would have still tried to engage at shorter ranges than this.

If the engagement would have followed the tendency of historical battleship engagements, i.e. decisive gun range would have corresponded to shorter ranges, it appears to be more likely that Littorio would have emerged victorious, but the zone of mutual immunity, especially at night, would have been more advantageous for Richelieu.

With her rate of fire increased to circa two rounds per gun per minute, the French ship could deliver almost thrice the amount of explosive as a function of time as compared to Littorio, and more accurately too, by virtue of her guns being radar-directed and their pattern size diminished. The mutual vulnerability of the main armament, conning tower and unarmored parts would have been, therefore, in Richelieu's favor. For instance, in case of a night engagement fought at 18-24 kyd (16.5-21.9 km)/60 deg. target angle, both ships' citadels would have been theoretically immune, but Richelieu could have most likely neutralized Littorio before the Italian ship could have done the same thing to her.

However, still assuming 60 deg. target angle, Richelieu would have had to sail within approximately 13 kyd (11.9 km) of Littorio to defeat her vertical armour, and even at such ranges, there would have been a high degree of uncertainty as to her shells being able to achieve this in operational condition. Also, this would have been within the range of Littorio's long-range AA guns that could fire starshells and the Italian ship could also use her searchlights and rudimentary radars to detect Richelieu. Therefore, even though she would have been at a clear disadvantage as far as fire-control is concerned, Littorio would not have been entirely blind, and she could pierce Richelieu's citadel regardless of target angle, either through her belt or transverse bulkheads, within 16 kyd (14.6 km).

As Illustrated, this would have been a hard battle after Richelieu's refit as well, but Littorio's superiority up to medium range, at least in case of good visibility, seems to have increased. However, by remaining in the zone of mutual immunity, Richelieu might have been able to neutralize her foe, especially at night, even though sinking her might not have been possible, unless she ventured dangerously close or managed to score fatal underwater hits. The superiority of the French ship at extremely long ranges was now somewhat more practicable too, albeit not significantly.

Appendix A – General Characteristics

Table AA.1

Designation	Littorio	Richelieu
Laid Down	28 October 1934	22 October 1935
Launched	22 August 1937	17 January 1939
Commissioned	6 May 1940	1 April 1940
Displacement (Standard) – tons (m.t.)	40,723 (41,377)	37,250 (37,848)
Displacement (Full Load) – tons (m.t.)	45,237 (45,963)	43,992 (44,698)
Length (Overall) – ft (m)	779.9 (237.7)	813.2 (247.85)
Length (Waterline) – ft (m)	762.5 (232.4)	794.0 (242.0)
Beam – ft (m)	107.9 (32.9)	108.5 (33.08)
Draught (Full Load) – ft (m)	34.4 (10.5)	32.5 (9.90)
Boilers	8 x Yarrow	6 x Sural
Turbines	4 x Belluzzo	4 x Parson
Machinery Output (Designed) – shp	130,000	155,000
Machinery Output (Overload) – shp	139,561	179,000
Speed (Overload) – knots	31.20	32.63
Fuel Capacity – tons (m.t.)	3,937 (4,000)	5,773 (5,866)
Electric Plant Output – kW	6,800	9,000
Range – Nmi	4,580 at 16 knots	9,850 at 16 knots
Armament (as designed)	9 x 15"/50 (3x3) 381mm/50 Ansaldo 1934	8 x 14.96"/45.41 (2x4) 380mm/45 Mle 1935
	12 x 6"/55 (4x3) 152.4mm/55 Ansaldo 1934	9 x 6"/55.05 (3x3) 152.4mm/55 Mle 1930
	4 x 4.72"/40 (4x1) 120mm/40 Armstrong 1891-99	12 x 3.94"/45 (6x2) 100mm/45 Mle 1930
	12 x 3.54"/50 (12x1) 90mm/50 Ansaldo 1938-9	8 x 1.46"/50 (4x2) 37mm/50 Mle 1933
	20 x 1.46"/54 (8x2 + 4x1) 37mm/54 Breda 1938-9	20 x 0.52"/76 (4x4 + 2x2) 13.2mm/76 Mle 1929
	20 x 0.79"/65 (10x2) 20mm/65 Breda 1935	---
Armour – tons (m.t.)	13,553 (13,770)	15,792 (16,045)
Aircraft (as designed)	3 x Ro. 43	4 x Loire 130
Complement	1,866	1,569
Radar (domestic)	EC 3/ter 'Gufo'	DEM

Appendix B – Calculating Penetration Values

The mathematical formula we apply to calculate immunity zones is based on the following assumptions:
1) We assume that total kinetic energy required to defeat spaced assemblies equals the sum of the individual energies needed to defeat each individual element.
2) We assume that in cases of laminated armour effective thickness equals the thickness of the thickest plate plus 70% of the thickness of the thinner plates.
3) Based on empirical data extracted from ADM 281/31 and ADM 281/37, we assume that the ballistic limit velocity of armour plates increases in proportion to the weight of armour piercing caps in cases of decapped projectiles. We assume that projectiles are decapped by armour plates thicker than 0.15 cal.

The relevant passages of ADM 281/37, which specifically deals with major caliber projectiles, is cited below.

1. INTRODUCTION
…It was decided to investigate the performance of decapped shell against various types of deck armour. For this purpose, 200 lbs./sq. ft. and 240 lbs./sq. ft. plates were ordered.
2. OBJECTIVE OF TRIALS
This was: -
(a) To compare the ballistic performance of non-cemented deck armour under attack by 15" A.P.C. at 60° and 65° with the shell in the capped and decapped conditions…

6. ANALYSIS OF RESULTS
…Penetration velocities for the 200 lbs/sq. ft. plates are shown in Table 1.

	TABLE 1		
Angle of Attack	Capped shell Wt. 1938 lbs.	Decapped Shell	
		Wt. 1712 lbs.	Wt. 1938 lbs.
60°	980 ft/sec.	1160 ft/sec.	1090 ft/sec.
65°	1030 ft/sec.	1285 ft/sec.	1210 ft/sec.

The figures for the capped shell are the mean of the results obtained for Investigation No. 8. The decapped shell weighs 1712 lbs. and the figures given in this column are the mean of the results obtained in the present Investigation. The figures for the decapped shell weighing 1938 lbs. are obtained from the results for the 1712 lbs. shell by assuming that the energy required by the shell to achieve penetration is the same i.e. mv^2=constant.

There is a marked difference between the performance of the capped and decapped shells and this still exists for capped and decapped shells of the same weight. An explanation of this may be that the shoulder of the cap succeeds in biting into the plate, even at such oblique angles as 60° and 65°, and, even though the cap may break, the shell will have been given an angular velocity tending to bring it towards the plate normal. The decapped shell on the other hand strikes on the comparatively flat shell shoulder and it is not until the shoulder has dented the plate that the shell is given a turning moment. Hence the decapped shell has a greater tendency to ricochet.

Cracking occurred in the 200 lbs/sq. ft. plates when attacked by the decapped shell, illustrating the fact that cracking is not caused entirely by the cap of a shell.

Table 2 gives similar information for the 240 lbs/sq. ft. N.C. plates: -

TABLE 2			
Angle of Attack	Capped shell Wt. 1938 lbs.	Decapped Shell	
		Wt. 1712 lbs.	Wt. 1938 lbs.
60°	1225 ft/sec.	>1400 ft/sec.	>1320 ft/sec.

The figure for the capped shell is the standard velocity for the given plate thickness and attack. The trend is the same as for the 200 lbs/sq. ft. plates.

The results for the C. and F.H. plates are very similar, (except for plate 4191 which broke up) and a penetration limit velocity of 1260 ft/sec. is a good representative value. This is well below the value of >1400 ft./sec. for the N.C. plates…

7. CONCLUSION

…(3) The penetration limit velocity for the de-capped 15" A.P.C. shell is as much as 200 ft./sec. above that for the capped shell when attacking 200 and 240 lbs./sq. ft. plates at 65° and 60°…

Summary of results obtained with 15" A.P.C. decapped shell

PLATE NO.	TYPE	ANGLE OF ATTACK	PENETRATION VEL.	TYPE OF CRACKING
8432	200 lbs. N.C.	65°	<1226	Heavy
8433	200 lbs. N.C.	60°	1160	Heavy
8434	240 lbs. N.C.	60°	>1437	Slight
8438	240 lbs. N.C.	60°	1370	Slight
4187	200 lbs. N.C.	65°	>1343	Slight
4186.A	200 lbs. N.C.	60°	1155	Moderate
4188	240 lbs. F.H.	60°	1220	Slight
4189	240 lbs. F.H.	60°	<1276	Moderate
4190	240 lbs. C.	60°	1280	Slight
4191	240 lbs. C.	60°	-	Heavy (Broke up)

ADM 281/37 indicates that limit velocity increases – on average – in proportion to the weight of the cap:

Table AB.1

Penetration Capacity of Capped and Decapped Projectiles				
Target		Capped	Decapped	Decapped/Capped
Thickness	Obliquity			
lbs	deg.	fps	fps	%
200	60	980	1,090	111
200	65	1,030	1,210	117
240	60	1,225	1,320	108
Average (fps)		**1,078**	**1,207**	**112**

ADM 281/31 indicates the same tendency in cases of smaller shells attacking C armour plates at 30 deg.

4) Angle of attack can be calculated as follows:
Let the tangent to the trajectory of the projectile at the point of impact and the plate normal line through the point of impact be represented by unit vectors in a three-dimensional Cartesian coordinate system. Now, calculate the angle between the two vectors.
5) Finally, we apply Bu. Ord. Sk. 78841.

Mathematically:

1) $VL_\Sigma = \sqrt{\sum_{i=1}^{n} VL^2_i}$

2) $t_{eff} = t_{max} + \sum_{i=1}^{n} 0.7 t_i$

3) $VL \propto \dfrac{m}{m - m_c}$

4) $\vec{t} = \begin{bmatrix} 0 \\ \cos(\alpha) \\ \sin(\alpha) \end{bmatrix}$

$\vec{n} = \begin{bmatrix} \cos(\gamma)\sin(\beta) \\ \cos(\gamma)\cos(\beta) \\ \sin(\gamma) \end{bmatrix}$

$\Theta = \cos^{-1}\left(\dfrac{\vec{t} \times \vec{n}}{|\vec{t}| \times |\vec{n}|}\right)$

5) $VL = \dfrac{F\sqrt{t}d}{41.57\sqrt{m}\cos(\Theta)}$

Whence
- VL=Ballistic Limit Velocity.
- VL_Σ=Total Ballistic Limit Velocity.
- m=Projectile Weight.
- m_c=Armour Piercing Cap Weight.
- t=Plate Thickness or Unit Vector (representing the tangent to the trajectory of the projectile at the point of impact).
- n=Unit Vector (representing the plate normal line through the point of impact).
- t_{max}=Thickness of Thickest Plate (laminated assemblies).
- t_{eff}=Effective Thickness.
- d=Projectile Diameter.
- Θ=Impact Obliquity.
- α=Angle of Descent.
- β=Lateral Angle.
- γ=Inclination of Armour Plate from the Vertical.
- $F=6*(t/d-0.45)*(\Theta^2+2{,}000)+40{,}000$.

Appendix C – Decapping Plates

The British Admiralty conducted a series of experiments that are very helpful to highlight the strengths and weaknesses of spaced assemblies and the phenomenon of decapping in general. The relevant parts of the *Admiralty Armour Investigation Program (1946-1950)* document ADM 281/31 are cited below.

1. INTRODUCTION

…In 1908 trials against an Italian assembly of: -

(a) A chrome-nickel plate 1.97" thick

(b) Wood backing 4.33" thick

(c) A Vickers K.C. plate 7.48" thick

showed that although there might be some merit in the system for sub-calibre attack (i.e. diameter of shell less than thickness of plate) there was no advantage against calibre or super-calibre attack.

In 1919, a spaced assembly, made up of a 180 lbs. plate separated by an 18" space from a 60 lbs. deflection plate, was subjected to attack by 15" A.P.C. shell at 60°, but gave disappointing results.

In 1937 reports were received that spaced armour was being fitted in the Italian battleship VITTORIO VENETO. A target was built up consisting of a 480 lbs. C plate secured to 80 lbs. N.C. plate by special armour bolts and this was attacked with 15" A.P.C. Mark XIII.A shell at 30° at 1613 ft./sec., the approximate limit velocity for a 600 lbs C plate. The assembly was defeated, the shell passing through with considerable remaining velocity.

A suggestion was made in 1943 by R.O.F. Cardonald that greater resistance to capped shell would be achieved by the use of two cemented plates together, with a thickness ratio of 1:1 or 1:2. About the same time it was learnt that the Italian battleship LITTORIO was fitted with a spaced assembly of: -

Front	*2.75" Krupp cemented armour*
Spacer	*6-7" "egg-box" construction*
Back	*11" Krupp cemented armour"*

Author's note: Apparently, British intelligence was not entirely accurate. The outer plate was OD, not KC; the spacer was about 10" (25 cm) wide and filled with water-repellent material or cement foam. Both plates had a thin ER backing plate and both were inclined about 15 deg. from the vertical (not 8 deg., as often quoted). The system was also reinforced by two inclined splinter bulkheads behind the KC plate and the upper strake of the innermost vertical torpedo bulkhead.

Trials were therefore arranged in 1945 to investigate these two types of assemblies…

It was shown that the spaced assembly had a slight advantage over the solid plate. It was decided, therefore, to investigate further and plates were ordered in the experimental program.

2. OBJECT OF TRIALS
This was to determine, using spaced assemblies of equivalent 320 lbs. thickness: -
(1) The most effective distribution of thickness
(2) The optimum U.T.S. of the front plate
(3) The most effective spacing of the front and back plates
(4) The relative advantage of using C. or NC. armour for the front plate.
(5) Whether the best spaced armour assembly was superior to a 320 lbs. C. plate.

...The effect of variation of the width of the "spacer" was not investigated and it was decided to use 10" spacer for all the trials... The spacer is assumed equivalent to 20 lbs. of plate thickness.

3. COMPOSITION AND MECHANICAL PROPERTIES OF PLATES
...The U.T.S. of all the back plates was about 50 tons/ sq. in., while for the front plates three ranges of U.T.S. values were considered, viz: 60 tons/ sq. in., 80 tons/ sq. in., and 100 tons/ sq. in.

The composition of the back plates was nominally the same but the front plates varied from 4% Ni., 2% Cr., to a 0.3% Ni., 3.4% Cr. composition.

4. DETAILS OF SHELL AND ATTACK
A preliminary trial was carried out in July 1948 to determine the best attack. Two shells were used, a 14" A.P.C. at 45° against an 80 lb./220 lb. assembly and an 8" S.A.P. at 30° against a 60 lb./240 lb. assembly.

...it was decided to use an attack intermediate between these two, and 9.2" A.P.C. shell was chosen at an angle of 30° to the normal.

5. CONDUCT OF TRIALS
The plates were set up as shown...This was found to be quite satisfactory.

6. ANALYSIS OF RESULTS
...Every shell fired against spaced assemblies in the 1949 trials was broken up but perforation of the rear plate was achieved at different velocities.

In the 1949 series the 60/240 lb. assembly was very inferior to the 80/220 lb. series. If the 60/240 lb. assembly of 1945 series is taken in conjunction with the 80/220 lb. and 40/260 lb. assemblies of the 1949 series a progressive improvement is seen as the thickness of the front plate is reduced. There is nothing in the mechanical properties of the 60/240 lb. assembly of the 1949 series to account for the very low perforation velocity obtained but there is no real justification for ignoring this result. The sequence however, seems a logical one, for all that is required of the front plate is to break up the cap of the shell and provided the front plate is thick enough to do this then the thicker the back plate the more effective will be the assembly in preventing complete perforation.

The A.R.E. report "The Best Use of Armour" states – "Experiments using flash photography show that the armour piercing cap is removed from A.P.C.B.C. shot by impact with thin mild steel or armour plate targets at all velocities above a critical velocity. The critical velocity varies with the hardness and thickness of the target and angle of attack and series of photographs have been obtained with increasing *striking velocity showing first loosening of the cap, displacement of the cap whole, nibbling of the cap at the rear edge, fracture of the cap at its critical velocity into a few larger pieces and then, at higher velocities, breakage of the cap into larger number of more widely dispersed fragments. The removal of an armour piercing cap by a thin frontal plate has an obvious application in the arrangement of divided armour, since, with the cap removed, the shot will be subject to shatter on a face hardened plate, the type of failure the A.P.C.B.C. shot is designed to avoid."*

The 40 lb. front plate breaks the cap as required and the results show that this 40/260 lb. assembly requires the highest velocity for perforation. As no further trials with thinner front plates have been carried out it must be concluded that, with a total thickness of 320 lbs, the best spaced assembly is 40/260 lbs.

There is no great difference between the results for the various 40/260 lbs. assemblies with various U.T.S. values for the front face. Thus, provided the front plate is hard enough to decap the shell there is no advantage to be gained by increasing its U.T.S. value. Similarly no advantage is gained by using cemented front plate and the trials with the 80 lb. C./220 lb. C. assembly confirms this when considered against the 80 lb. N.C./220 lb. C. assembly.

The results for the single 320 lb. C. plates... of 1945 was considered a bad basis for comparison, and another plate... was tested in 1949. A shell velocity of 1770 ft./sec. was stopped by the plate, a non-penetration velocity higher than for any of the spaced assemblies. Perforation without shell break up however, occurred at a velocity similar to the perforation velocity for the best assembly which did effect break up. From this point of view the spaced assembly is superior as effective shell break up occurs at velocities where perforation without shell break up would be sustained by the solid plate.

The 1948 trials, ... which were carried out to determine the best form of attack in the subsequent trials, showed that against super calibre attack the spaced assembly is not efficient. The attack was 14" A.P.C. against an 80/220 lb. assembly and perforation was obtained at velocities as low as 1113 ft./sec. whereas the perforation limit for 320 lb. C. armour against this attack is of the order of 1250-1300 ft./sec.

7. CONCLUSIONS
Based on trials with attack by 9.2" A.P.C. shell at 30° against assemblies with an equivalent weight of 320 lbs./sq. ft.: -

(1) Against calibre or sub calibre attack spaced armour is nearly 100% effective in breaking up capped shell.
(2) The perforation limit for spaced assemblies against calibre or sub calibre attack is approximately the same as for a solid plate of the same total thickness.
(3) Against super calibre attack a spaced assembly is inferior as regards perforation velocity to a solid plate of the same overall poundage.
(4) For the type of attack and equivalent total thickness considered in this report the most effective thickness of front plate is 40 lbs. and an N.C. plate of U.T.S 60 tons/ sq.in. is as effective as a C. plate.
(5) The trials do not enable conclusion to be reached regarding the most effective width of spacer. In the case of ship side armour the scope for variation is small and a spacer width approximately of the same order as the shell calibre is unlikely to be exceeded. In these conditions it is considered unlikely that the width of spacer will have an appreciable effect on the results.

Appendix D – Richelieu's Secondary Armament Fire-Control System

Table AE.1

3 x Secondary Armament Directors			
Télépointeur 2 (Forward middle director)		Télépointeur 3 (Aft director)	
Director Optics)			
Rangefinder)	1 x OPL	Rangefinder)	1 x OPL
Type	Stereoscopic	Type	Stereoscopic
Base length	26.25 ft. (8.0 m)	Base length	19.69 ft. (6 m)
Director Radar) (From 1945)		Director Radar) (From 1945)	
Fire Control Radar	1 x Type 285P (UK)	Fire Control Radar	1 x Type 285P (UK)
Range)			
Battleship	15000 m (16404 yds.)		
Accuracy)			
Range	49 yds m (45 m)		
Bearing	0.25°		
Power	150 kW		
Wavelength	50 cm		
Frequency	600 MHz		
Resolution	164 yds (150 m)/1.5°		
Télépointeur 1 (Forward uppermost director) (Removed in 1943)		3 x Secondary turrets	
Director Optics)		Turret Optics)	
Rangefinder)	1 x OPL	Rangefinder)	1 x OPL
Type	Stereoscopic	Type	Stereoscopic
Base length	19.69 ft. (6 m)	Base length	26.25 ft. (8.0 m)

Appendix E – Jean Bart's Guns and Directors

Figure AE.1

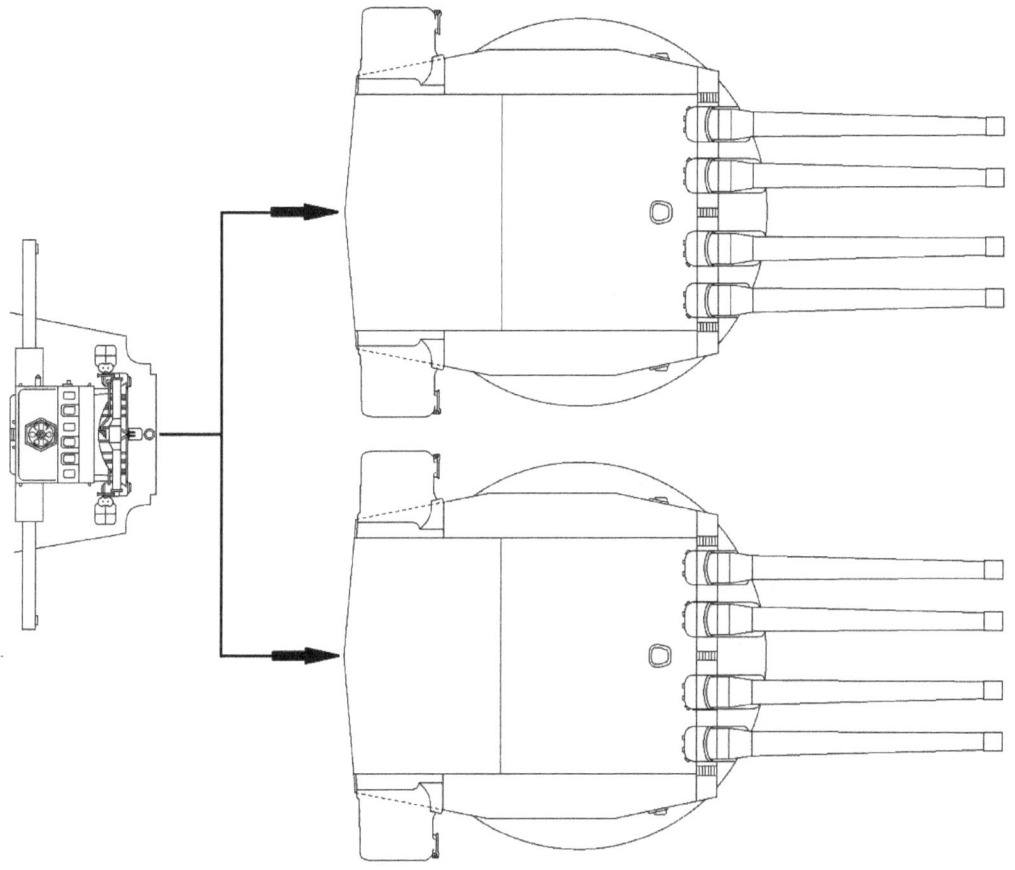

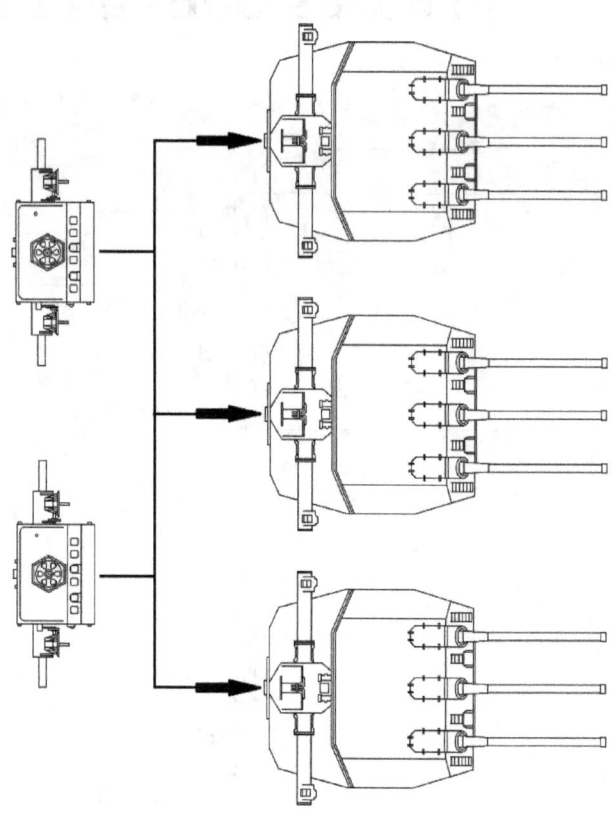

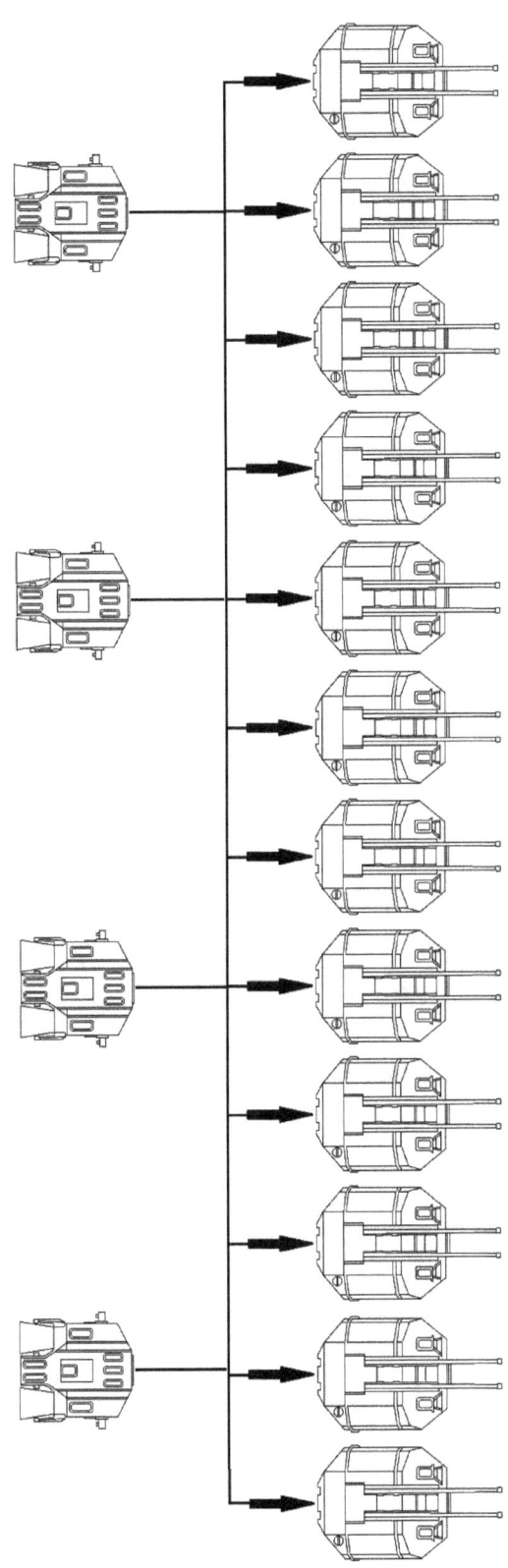

Figure AE.2

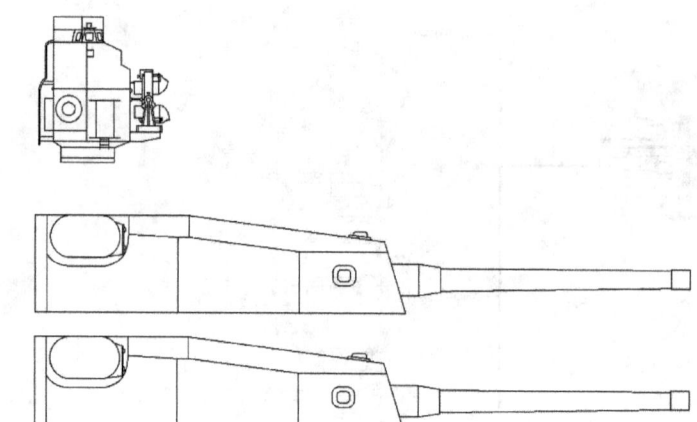

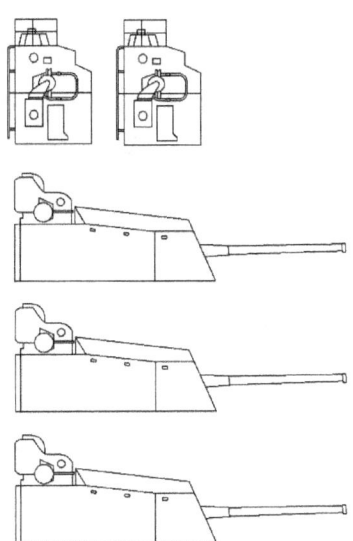

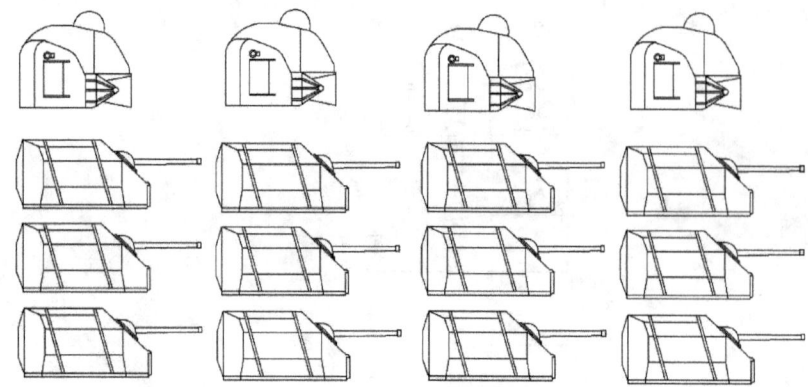

Appendix F – U.S. Radars of Richelieu

Figure AF.1

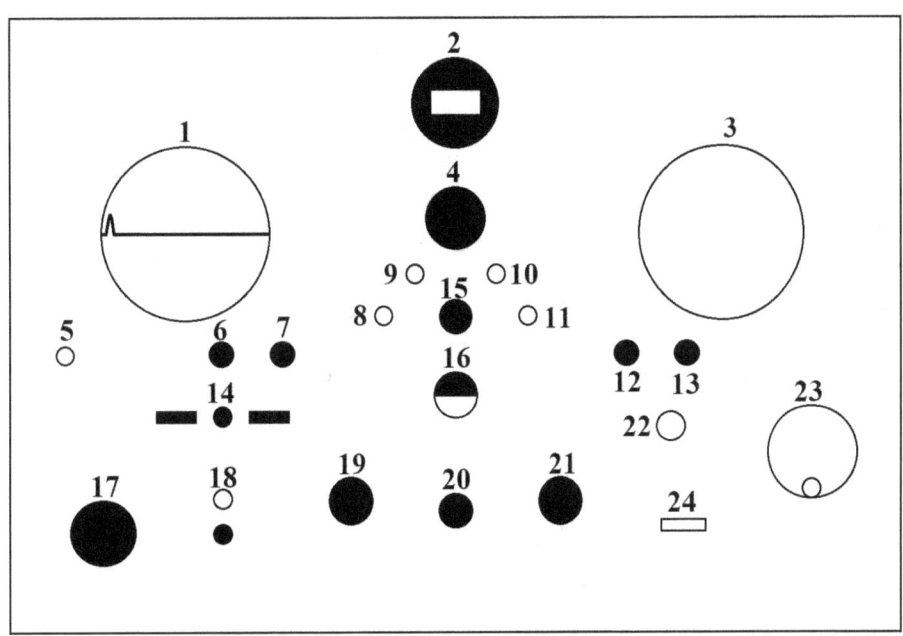

	SF-1. Indicator unit.		
1	"A" indicator	13	PPI intensity
2	Range scale	14	Calibrate-operate switch
3	PPI indicator	15	Dial light control
4	Range knob	16	Green tuning eye
5	Cal synch	17	Rec-gain control
6	"A" scope intensity	18	Stop-start buttons
7	"A" scope focus	19	Range switch
8	16,000-yard set	20	Lo-tuning
9	16,000-yard range set	21	IFF gain
10	48,000-yard range set	22	Warning-training error
11	48,000-yard zero set	23	Antenna train control
12	PPI focus	24	IFF on-off switch

Figure AF.2

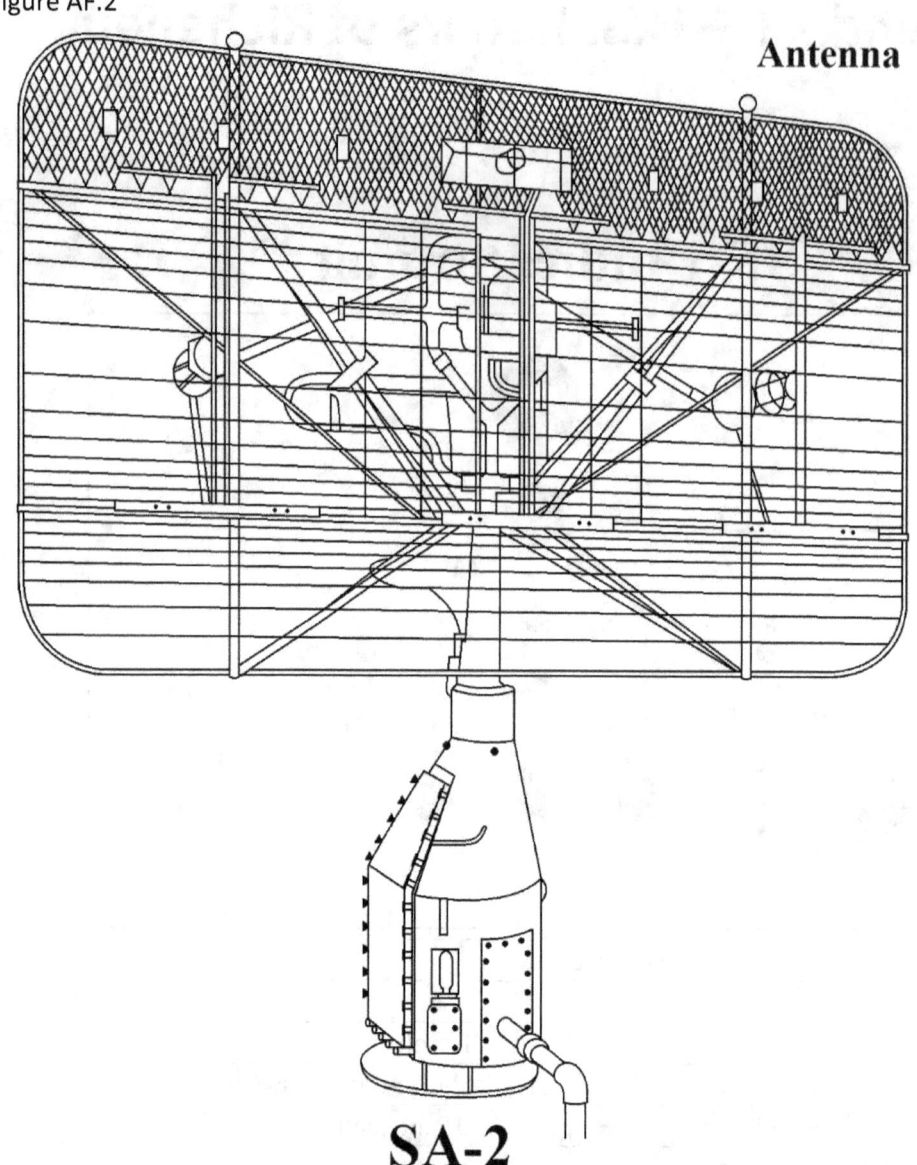

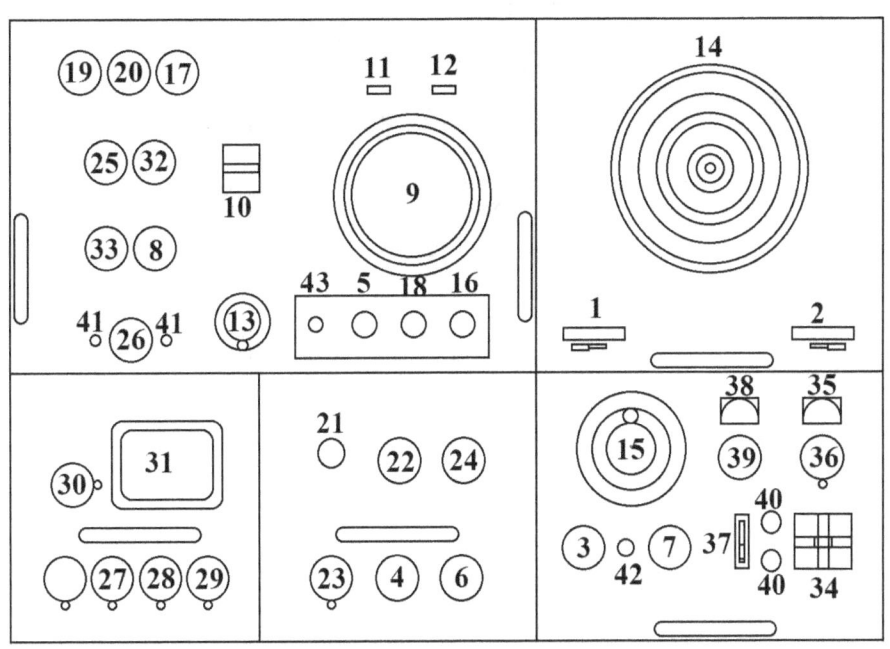

Receiver indicator unit

	Receiver Indicator Unit		
1	Antenna train-relative, true	23	Cal. osc.
2	Emergency train	24	Cal. max.
3	Slewing motor-low, off, high	25	Cal. min.
4	Range sel. (range selector)	26	Dimmer
5	Cal-fid IFF	27	First R.F.
6	Cal synch (calibrate, synchronizing switch)	28	Second R.F.
7	L.R. motor-on, off, on	29	Osc.
8	L.R. off	30	Ant.
9	Range oscilloscope (the scope)	31	Calibration chart
10	Yards range counter	32	IFF gain
11	"B" miles range counter	33	L.R. amp
12	"C" miles range counter	34	Main power switch
13	Range step control	35	Line voltage meter
14	Bearing indicator	36	Line voltage variance
15	Manual antenna-train knob	37	Transmitter power switch
16	Focus	38	Plate current meter
17	Astig (Astigmatism control)	39	Plate voltage variance
18	Intensity	40	Fuses
19	Horiz (Horizontal centering control)	41	Fuse
20	Vertical (Vertical centering control)	42	Bearing mark
21	Oscillator adjustment oscilloscope	43	Range mark
22	Oscillator adjustment oscilloscope focus		

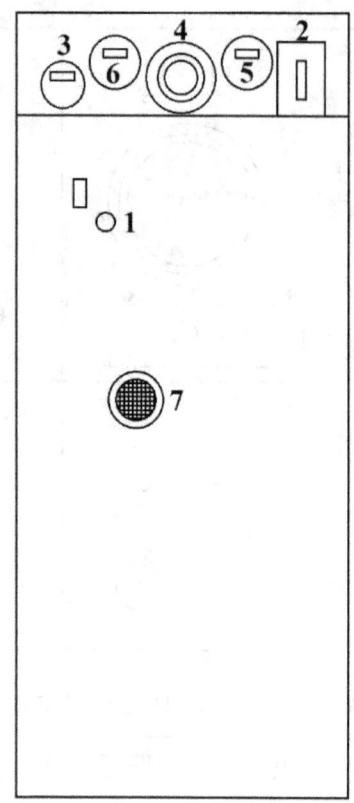

Transmitter unit

	SA-1. Transmitter unit
1	Duplexer adjustment
2	Emergency off
3	Plate current meter
4	Fil. primary voltage variance
5	Fil. voltage meter
6	Hours operation meter
7	Blower opening

Figure AF.3

SG Range and train indicator unit

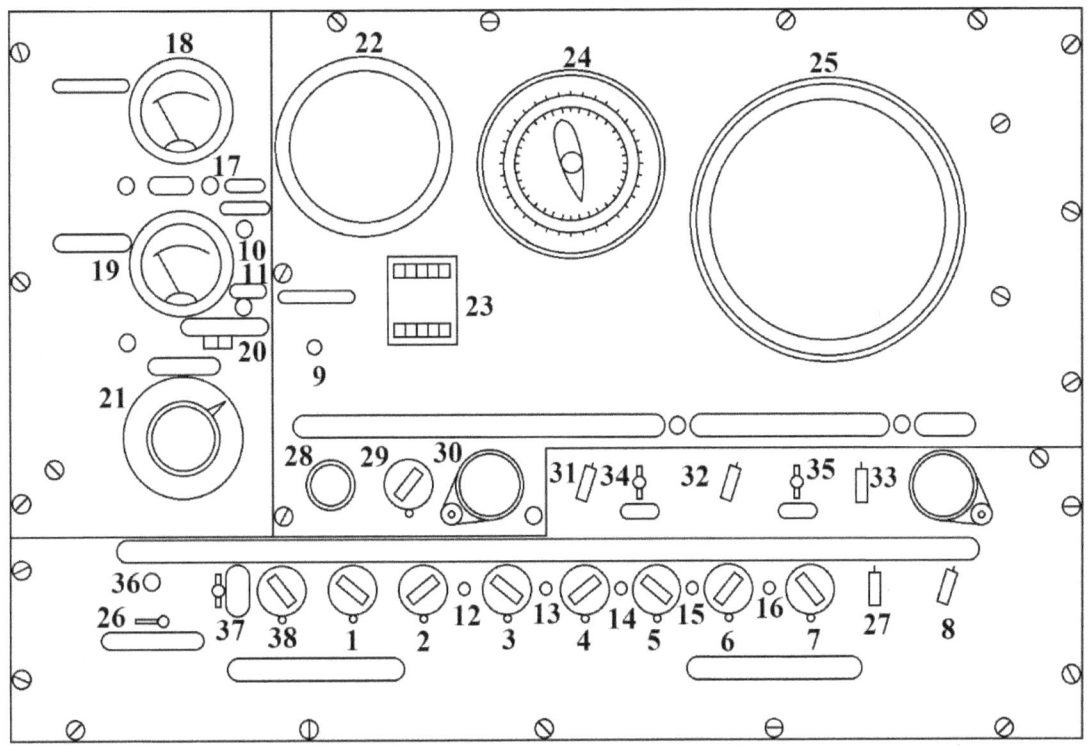

	SG – Range and train indicator unit.		
1	Range focus	20	Radiation switch
2	Zero set	21	Power supply control
3	Limit set	22	Range scope
4	Zero set	23	Range counters
5	Limit set	24	Bearing indicator
6	Pulse repetition frequency control	25	PPI
7	PPI focus	26	Synchro switch
8	Dial lights switch	27	When the radar is operating, switch K is in the NORMAL position
9	Pilot lights switch	28	Receiver gain control
10	Indicator F-902	29	Receiver tune control
11	Bearing Control F-901	30	Range crank
12	H center adjustment	31	Range scale switch
13	V center adjustment	32	Allows the operator to receive either signals or range markers on the range scope and PPI
14	Marker amplitude adjustment	33	Antenna's rotation control
15	PPI intensity adjustment	34	Remote range switch
16	PPI anode adjustment	35	Remote bearing switch
17	Remote control for the main-power switch at the transmitter-receiver unit	36	Reset button
18	Line voltage indicator	37	Determines the positions Off, Intermittent, and Continuous operation for IFF equipment
19	Transmitter current indicator	38	IFF gain adjustment

Appendix G – Projectiles

Figure AG.1

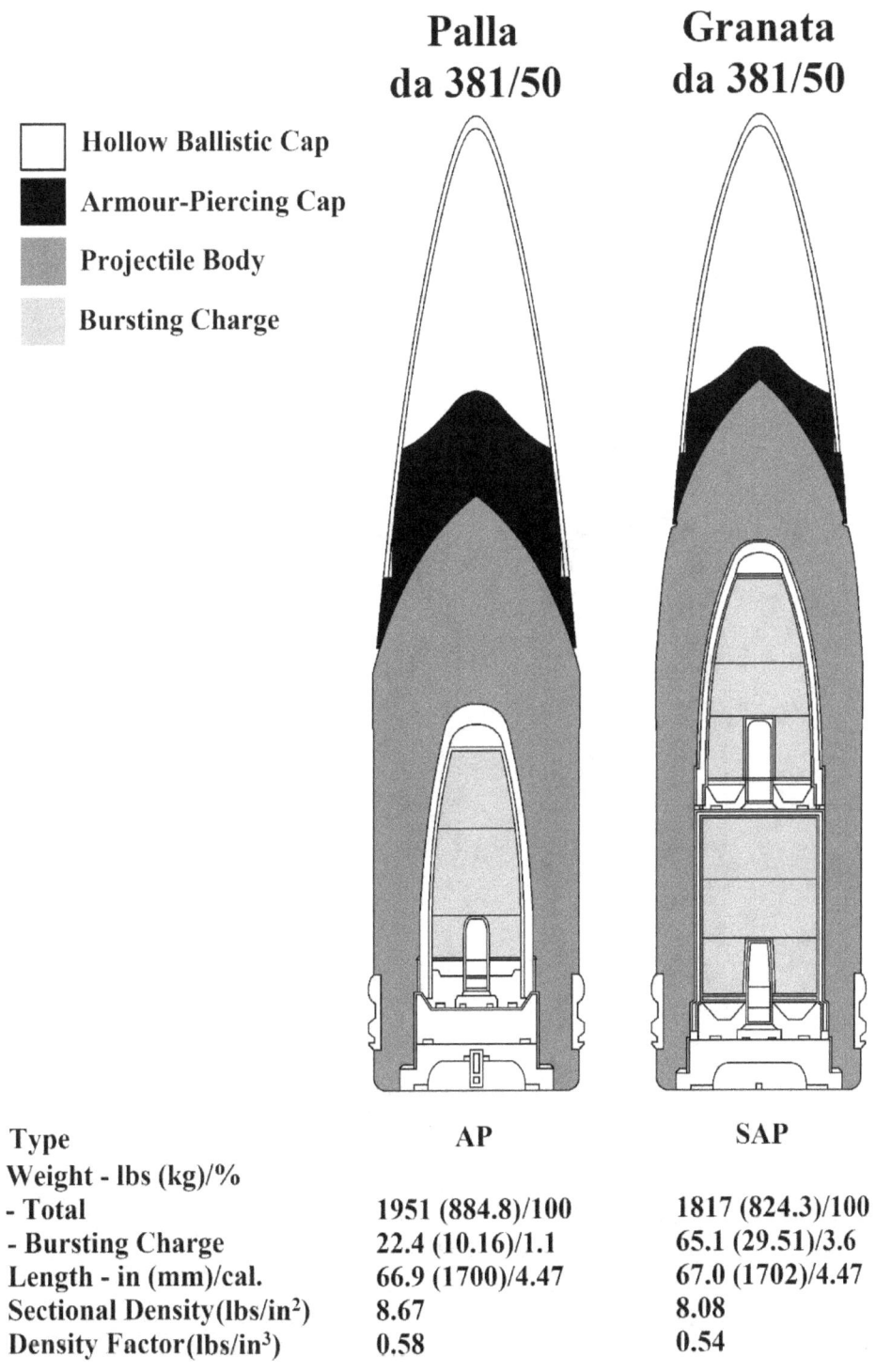

Type	AP	SAP
	Palla da 381/50	Granata da 381/50
Weight - lbs (kg)/%		
- Total	1951 (884.8)/100	1817 (824.3)/100
- Bursting Charge	22.4 (10.16)/1.1	65.1 (29.51)/3.6
Length - in (mm)/cal.	66.9 (1700)/4.47	67.0 (1702)/4.47
Sectional Density (lbs/in^2)	8.67	8.08
Density Factor (lbs/in^3)	0.58	0.54

Figure AG.2

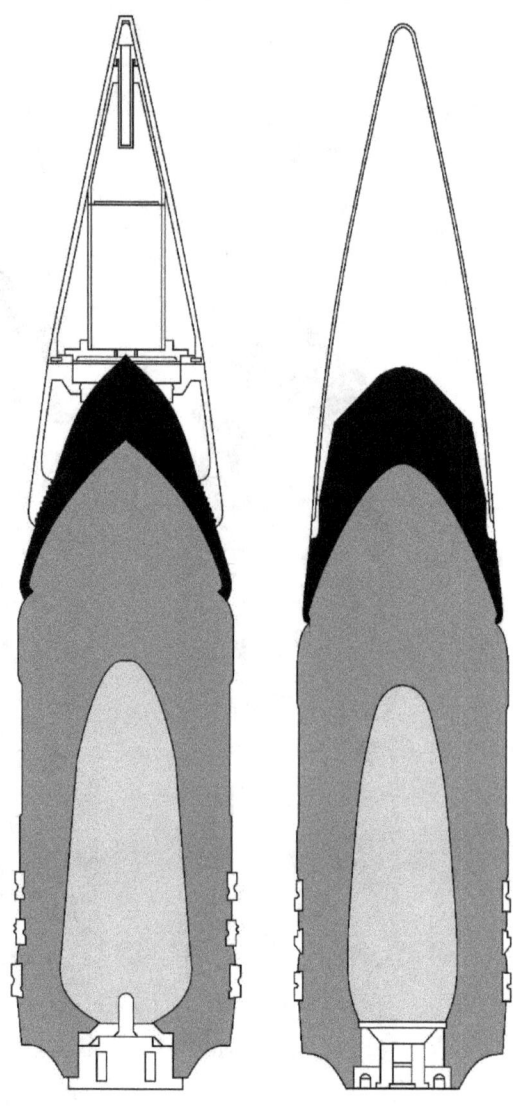

38 cm Projectiles
OPfK Modéle 1936   1943

- ☐ Hollow Ballistic Cap
- ■ Armour-Piercing Cap
- ▨ Projectile Body
- ▨ Bursting Charge

Weight - lbs (kg)/%	1936	1943
- Total	1949/884/100	1949/884/100
- Armour-Piercing Cap	185/84/9.5	276/125/14.0
- Bursting Charge	48.3/21.9/2.5	44.4/20.1/2.3
Length - in (mm)/cal.	74.8/1900/5	74.1/1882/4.95
Sectional Density (lbs/in²)	8.7	8.7
Density Factor (lbs/in³)	0.58	0.58

Appendix H – Jean Bart's Protection Scheme

Figure AH.1

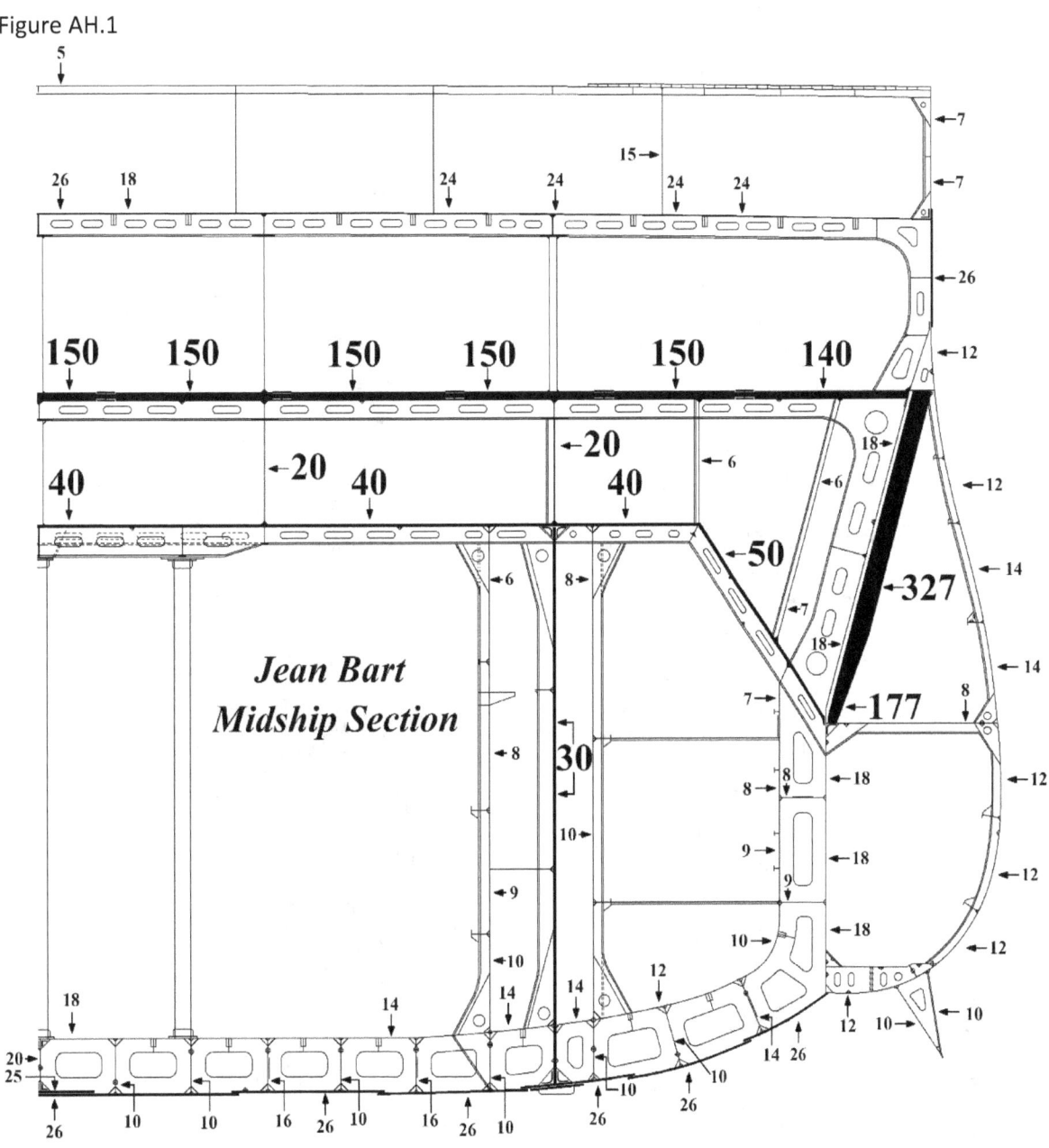

Longitudinal Section

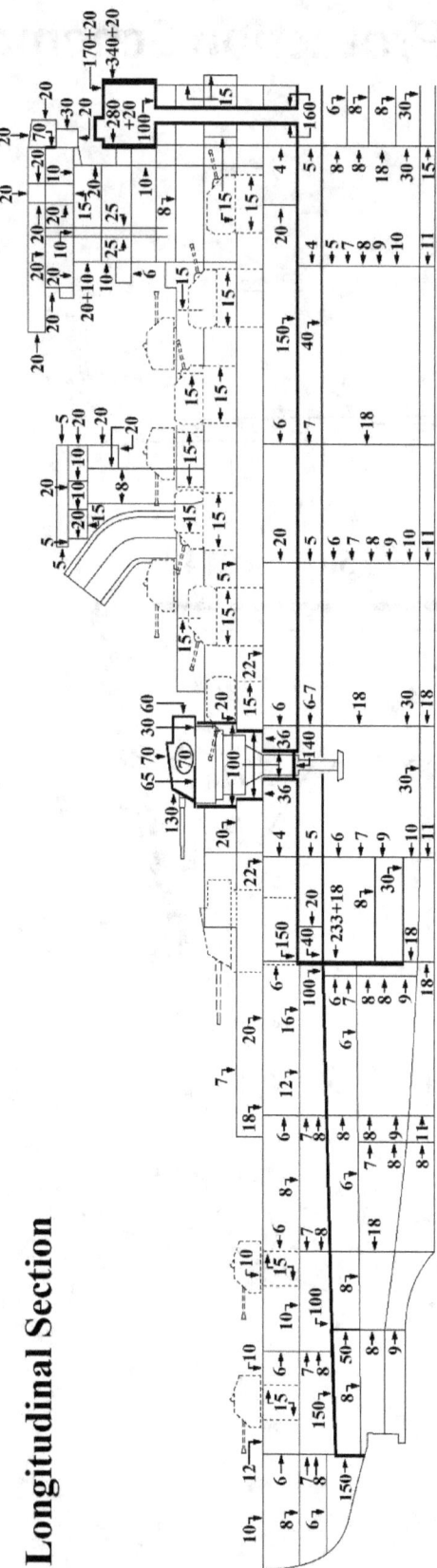
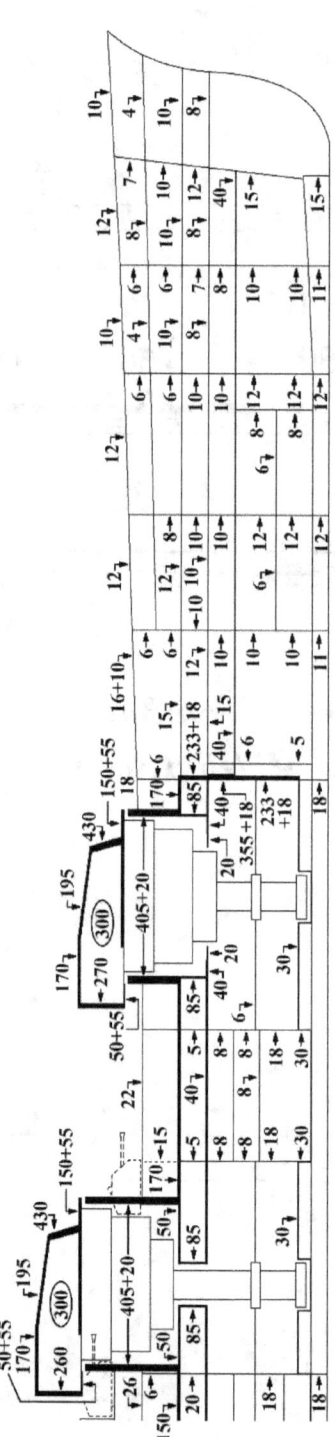

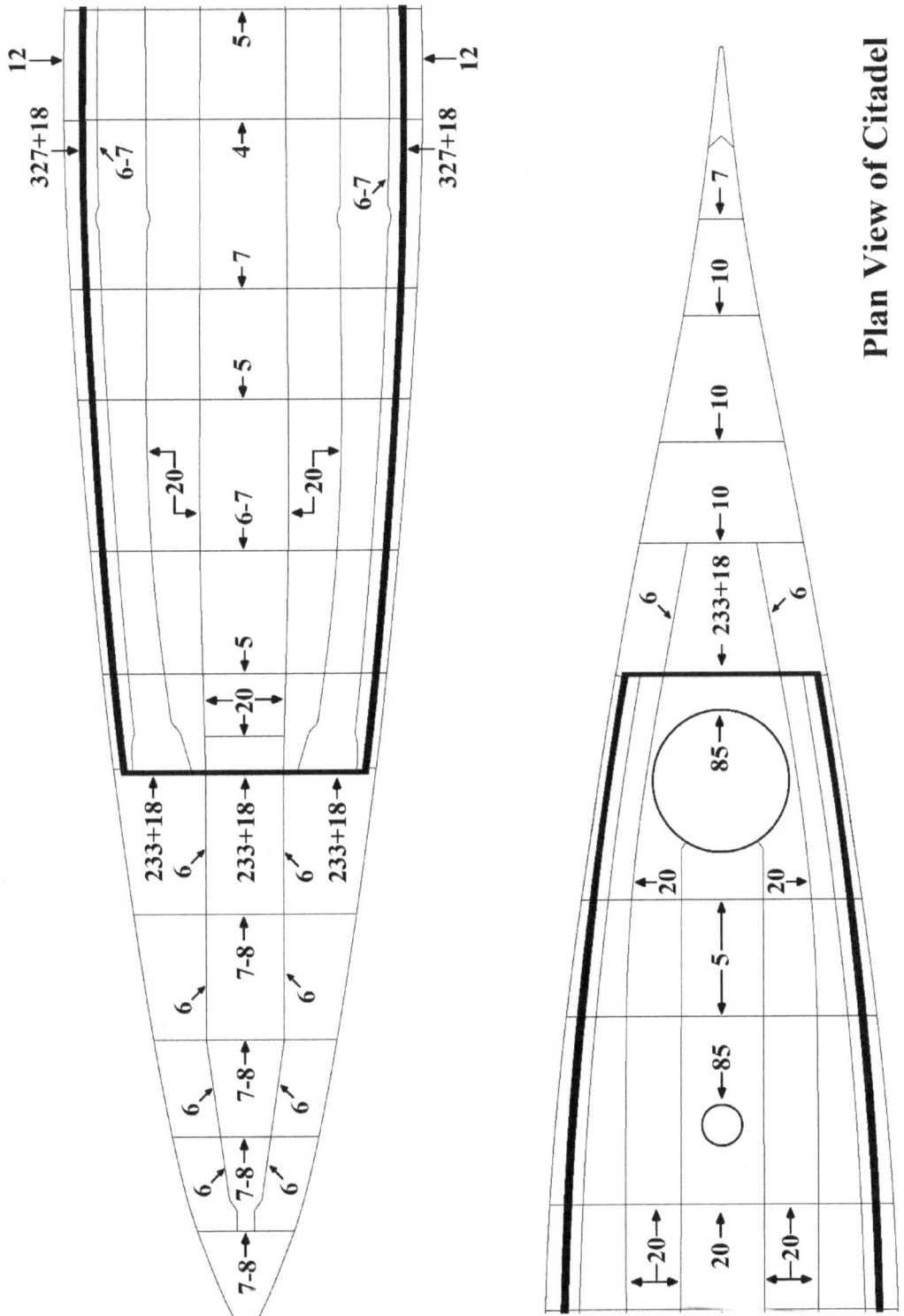

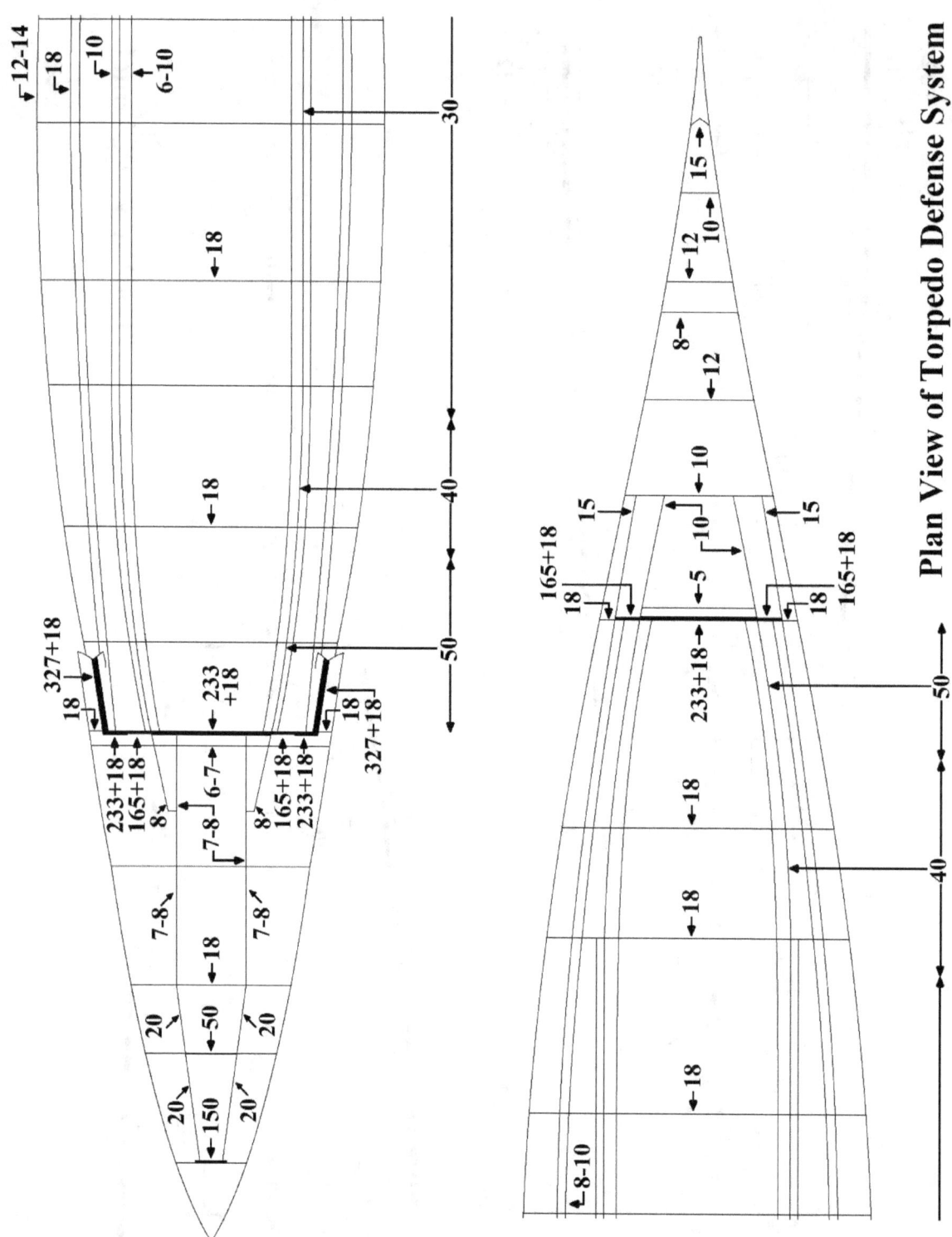

Plan View of Torpedo Defense System

Transverse Bulkheads

Citadel

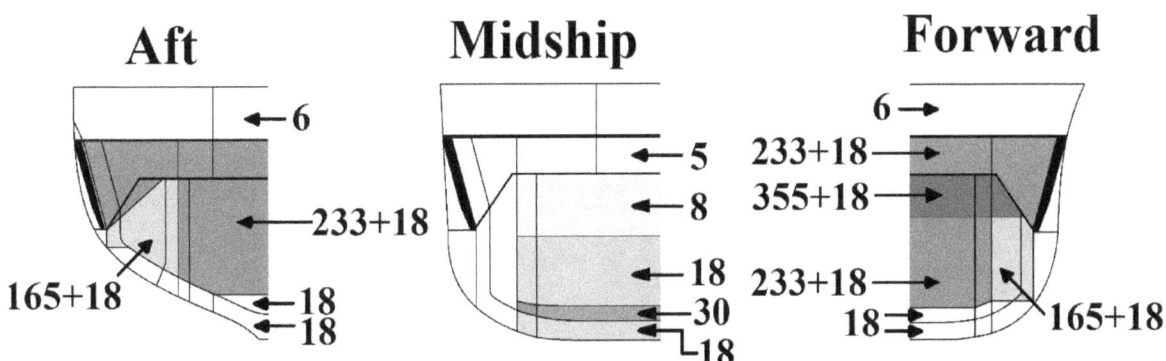

Steering Gear Room

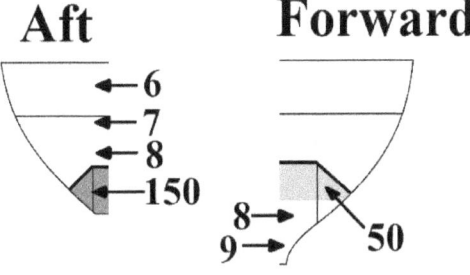

Appendix I – Turret Mounts

Figure AI.1

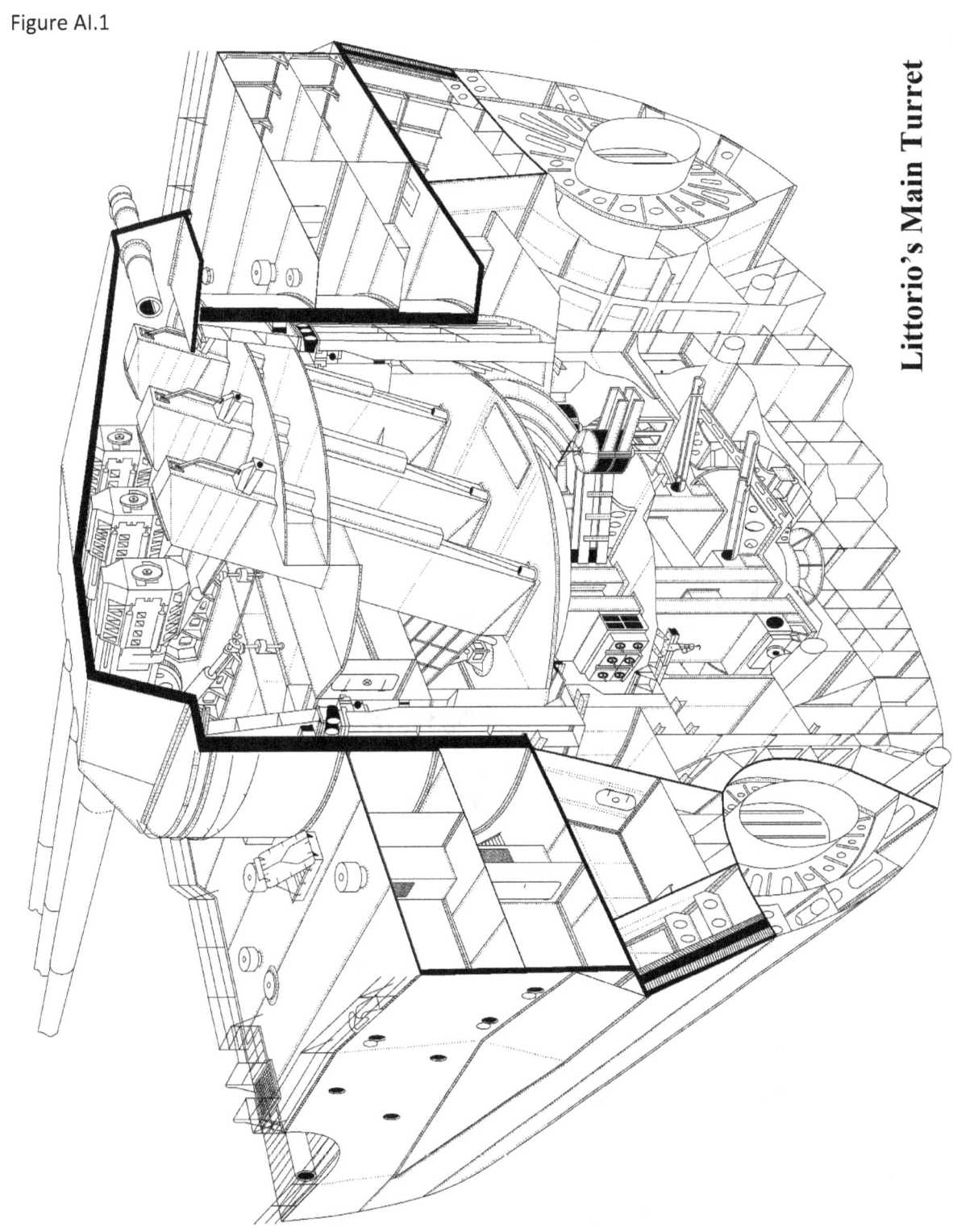

Littorio's Main Turret

Figure AI.2

Richelieu's Main Turret

Appendix J – Longitudinal Sections

Figure AJ.1

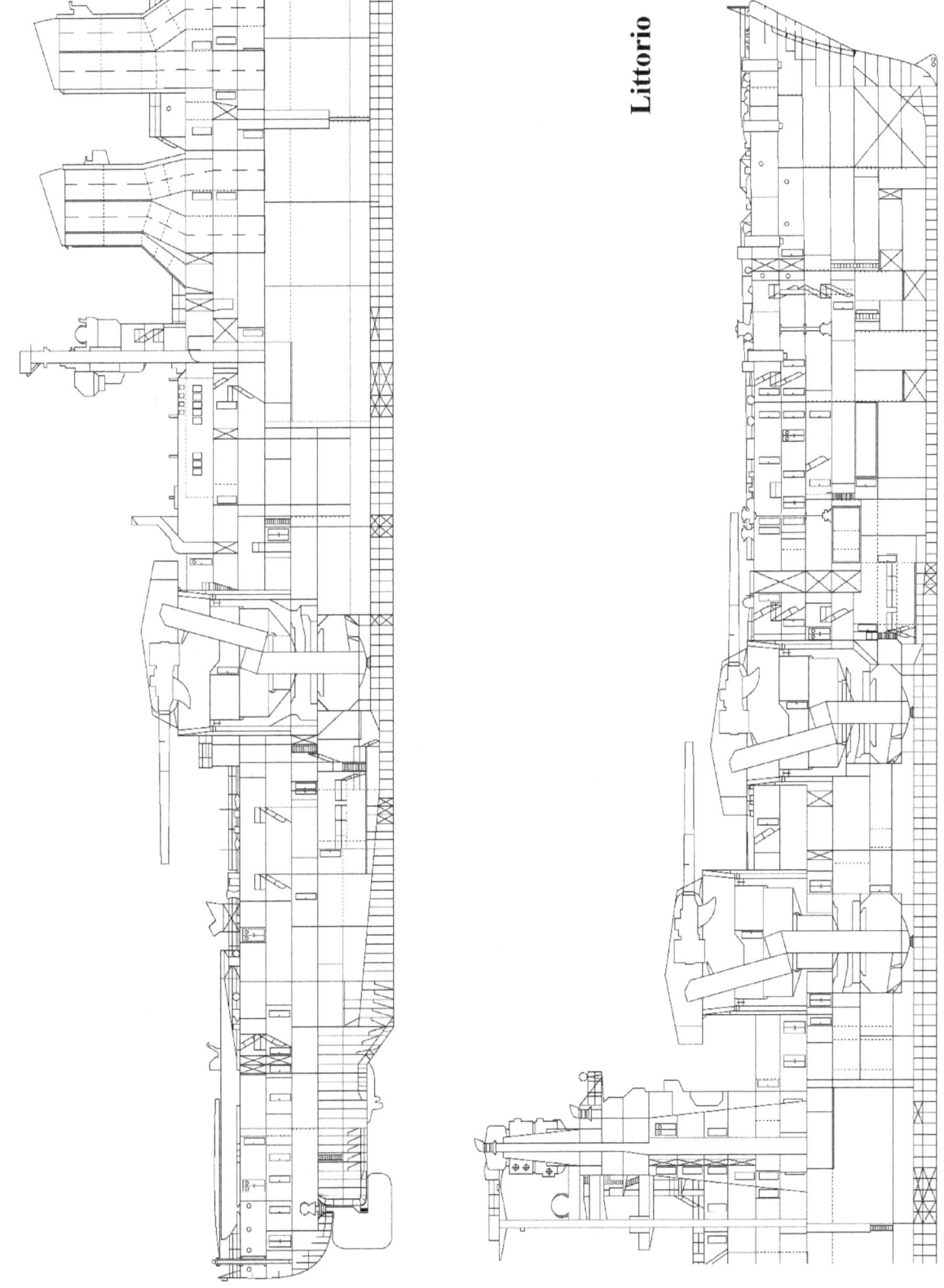

Figure AJ.2

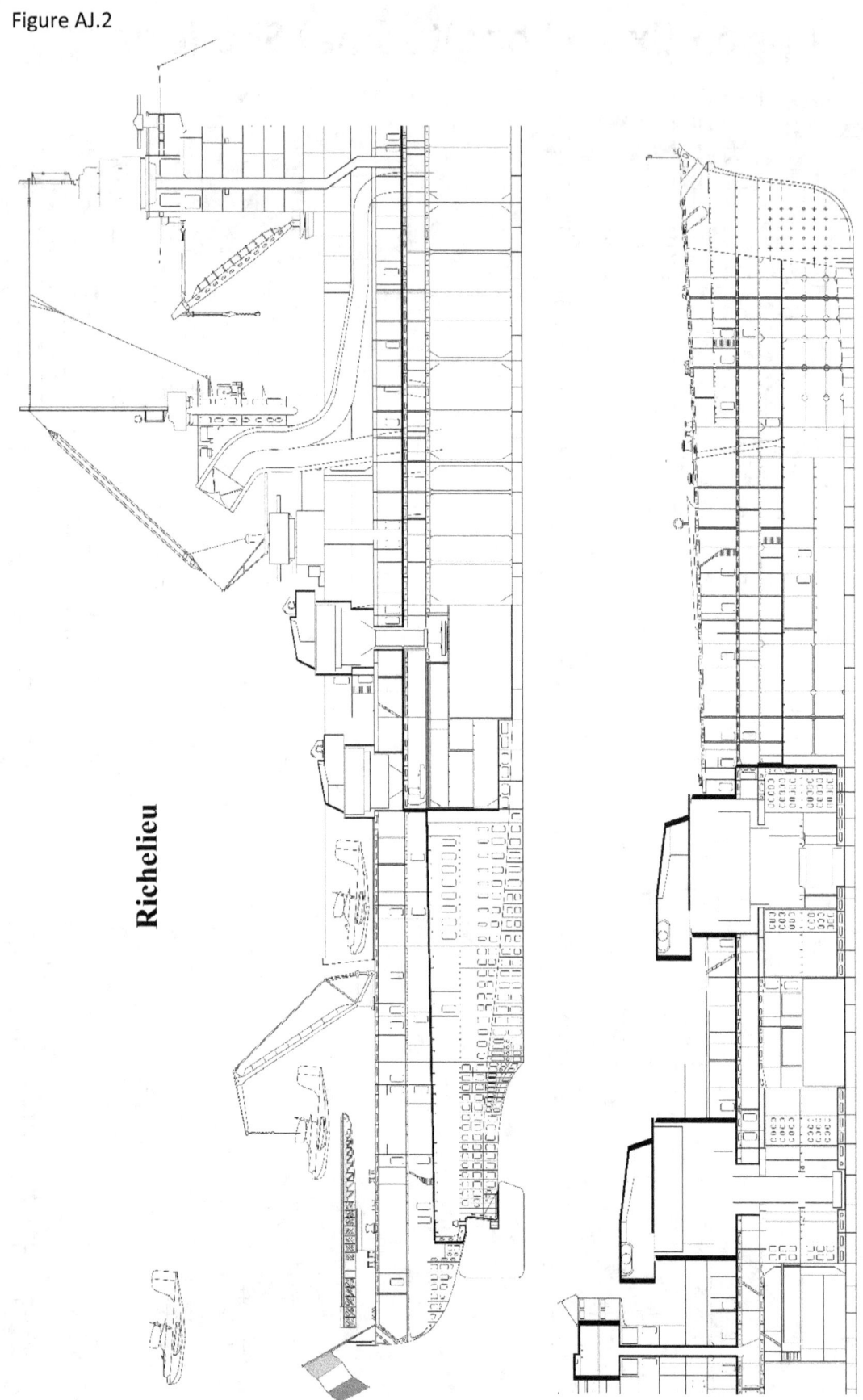

Richelieu

Appendix K – Major Caliber Hits Scored Against French Capital Ships

Table AK.1

Hit	Target	Thickness (in)	VL based on Bu. Ord. Sk. 78841 (fps)	Result
Major Caliber Hits Scored Against French Capital Ships				
Jean Bart Vs. U.S. 16in/2,700 lbs Shells (velocity:~1,480 fps/descent:~25 deg.) at Casablanca)				
No. 1	**Armour Deck** **Splinter Deck**	5.91 1.57	**1,710**	Penetrated
No. 2	Unarmored Region	---	---	---
No. 3	Unarmored Region	---	---	---
No. 4	Main Belt	12.87 at 15.4 deg.	1,670*	Ricocheted
No. 5	Main Turret Glacis	5.91+2.17	1,935	Ricocheted
No. 6	Main Turret Barbette	15.94	1,590*	Ricocheted
No. 7	Sloped Deck Aft	3.94 at 45 deg.	580*	Penetrated
Richelieu Vs. British 15in/1,938 lbs Shell (velocity:~1,550 fps/ descent:~19 deg.) at Dakar)				
No. 1	Armour Deck	5.91	2,300	Ricocheted
Dunkerque Vs. British 15in/1,938 lbs Shells (velocity:~1,700 fps/ descent:~12 deg.) at Mers el Kébir)				
No. 1	Turret Roof (Fwd.)	5.91 at 9 deg.	2,100*	Ricocheted
No. 2	Unarmored Region	---	---	---
No. 3	Main Belt	8.86 at 11.3 deg.	1,120*	Penetrated
No. 4	Main Belt Slope T. Bhd.	8.86 at 11.3 deg. 1.57 at 34 deg. 1.18	1,260*	Penetrated
*=at 0 deg. lateral angle. Lateral angle was far greater than this at Casablanca and slightly greater than this at Mers el Kébir. Therefore, limit velocity was far higher at Casablanca and slightly higher at Mers el Kébir than the values given.				

Appendix L – Official Penetration Capacity and Gunnery Trials of Littorio's Main Guns

Table AL.1

Armour Penetration Capacity of Littorio – in (mm)						
Range	Vertical Armour at 90 deg.		Vertical Armour at 60 deg.		Horizontal Armour	
yds (m)	Official*	Bu. Ord. Sk. 78841	Official*	Bu. Ord. Sk. 78841	Official*	Bu. Ord. Sk. 78841
20,779 (19,000)	16.38 (416)	19.48 (495)	13.31 (338)	15.10 (384)	2.64 (67)	3.12 (79)
21,872 (20,000)	15.83 (402)	18.90 (480)	12.83 (326)	14.68 (373)	2.91 (74)	3.26 (83)
26,247 (24,000)	13.70 (348)	16.76 (426)	11.22 (285)	13.15 (334)	4.13 (105)	4.18 (106)
28,434 (26,000)	12.80 (325)	15.82 (402)	10.47 (266)	12.50 (318)	4.88 (124)	4.64 (118)
*=Direttive e norme per l'impiego della Squadra navale. Impiego delle artiglierie, September 1942.						

Table AL.2

Armour Penetration Capacity of Littorio – in (mm)				
Range	Vertical Armour at 90 deg.		Horizontal Armour	
yds (m)	O.T.O. Melara	Bu. Ord. Sk. 78841	O.T.O. Melara	Bu. Ord. Sk. 78841
0	32.05 (814)	32.09 (815)	---	---
19,685 (18,000)	20.08 (510)	20.06 (510)	2.87 (73)	2.93 (74)
30,621 (28,000)	14.96 (380)	14.95 (380)	5.12 (130)	5.13 (130)

Table AL.3

Gunnery Exercise Results (1938-1941)							
Source: G Colliva, 'Il trio navale italiano' in STORIA militare, n. 199							
Gun	Charge	Date	Range	Rate of fire	Deflection error	Range error	Hits
Manufacturer	No.	Year	Kyd (km)	sec	% of range		%
O.T.O. 381/50	1st	1939-40	23.0 (21.0)	63.0	1.38	1.27	7.0
	2nd	1939-40	18.6 (17.0)	37.5	1.09	2.02	10.7
	2nd	1940-41	21.9 (20.0)	29.7	2.11	2.50	7.3
Ansaldo 381/50	1st	1939-40	24.6 (22.5)	59.0	1.85	1.62	3.7
	2nd	1939-40	19.0 (17.4)	47.0	2.07	1.81	9.5
	2nd	1940-41	20.6 (18.8)	30.6	1.64	1.91	6.2
Average	---	---	21.3 (19.5)	44.5	1.70	1.84	7.4

Definitions

Angle of attack – The angle between the tangent to the trajectory of the projectile at the point of impact and the plate normal line through the point of impact.

Angle of descent – The vertical angle between the horizontal plane and the tangent to the trajectory of the projectile at the point of fall.

Angle of fall – Same as angle of descent.

Armour citadel – The passive protective system formed by the main armour belt, the main armoured transverse bulkheads and the main armour deck.

Armour penetration capacity – Capability to defeat armour.

Armour-piercing cap – Designed to enhance their armour penetration capacity and bite angle, armour-piercing projectiles are fitted with this kind of cap.

Armour-piercing projectile (AP) – Designed to defeat heavy armour, projectiles of this type have thick metallic walls and contain a comparatively limited amount of explosive.

Average gun – A used gun with a degree of bore enlargement corresponding to half the life of the barrel in terms of equivalent full charges (EFC) fired. Muzzle velocity, ballistic characteristics and vertical armour penetration capacity of an average gun compared to that of a new gun are inferior, but horizontal armour piercing capacity is increased due to more plunging trajectories.

Average pressure profile – Constant function having the same limit points and integral value as the function of the actual pressure inside the gun.

$\int p(x)dx = \int p_{av}(x)dx = p_{av} * L$

whence

p = pressure,

p_{av} = average pressure,

L = Length of gun barrel.

Ballistic cap – Streamlined, hollow windshield provided to reduce retardation.

Ballistic characteristics – Characteristics dependent on initial velocity and the ballistic coefficient of the projectile, namely max. range, time of flight, max. ordinate, angle of fall, drift, danger space etc. The flatter the trajectory the better the ballistic characteristics of the gun. The higher the initial velocity and the more favorable the ballistic coefficient of the projectile, the flatter the trajectory.

Ballistic coefficient – Dependent on the sectional density and shape of the projectile, ballistic coefficient indicates the ability of the projectile to overcome the resistance of the medium it travels through. Retardation greatly depends on this coefficient.

Ballistic limit velocity (VL) – Terminal velocity required for a given projectile to perforate a plate of given type and thickness at a specified obliquity.

Barrel life – The life of a gun barrel is measured by the number of equivalent full charges (EFC) that can be fired without having to reline the gun.

Boat tail – Taper of the projectile behind the driving band as an attempt to enhance its ballistic coefficient.

Bore erosion – Amortization of the gun bore due to usage. The higher the average pressure profile of the gun, the more severe the erosion.

Bursting charge – Explosive filling of projectiles.

Caliber radius head (crh) – Measurement of how streamlined the head of a projectile is. Larger numbers indicate more pointed shell heads.

Capital ship – Armoured vessel of war, not an aircraft carrier, mounting a battery of a caliber greater than 8 inches (203 mm).

Cemented armour (C) – Armour plates of this type have hardened face in order to shatter the armour-piercing cap/body tip of AP projectiles. The backing layer of the hardened face has comparable metallurgical properties to those of homogeneous armour plates. The two layers are often separated by a transitional layer, the hardness of which is gradually reduced near the backing layer. Cemented armour is most efficient when impact angles are relatively close to the normal. Hence, heavy vertical armour was usually constructed of this material.

Danger space – The distance measured along the line of fire in front of the target that, if the target were moved toward the firing point, a shot striking the base of the target in its original position would strike the top of the target in its new position.

Decapped projectile – Capped armour piercing projectile whose armour piercing cap has been knocked off or shattered.

Density factor – Projectile weight divided by the cube of bourrelet diameter (m/D^3).

Detrimental hardening – Projectiles having this type of hardness distribution have virtually uniform hardness on the surface and at the center. The British and the Japanese preferred this hardness distribution.

Director – Artillery position housing fire-control sensors.

Dispersion – The distance of the point of impact of the shot from the mean point of the impact of the salvo. Dispersion in range is measured along the line of fire and in deflection at right angles to the line of fire.

Drag – The effect of the resistance of the medium which the projectile travels through.

Firing delay – Deliberate pause of a fraction of a second between the firing of individual shots to prevent interference.

High explosive projectile (HE) – Designed to destroy lightly protected targets, projectiles of this type have relatively thin metallic walls and contain large quantities of explosive as compared to armour-piercing shells.

High-performance gun – Refers to guns with above-average muzzle energy for their respective bore diameter. High-performance guns have a high average pressure profile.

High-pressure gun – Refers to guns with an above-average average pressure profile. Guns having an average pressure profile of over 9.5 tons/in^2 are termed high-pressure guns in this study.

High-velocity gun – Guns with at least above-average muzzle velocity. Guns having an initial velocity of at least 2,600 fps (792 mps) are termed high-velocity guns in this study.

Homogeneous armour – Armour plates of this type have theoretically uniform mechanical properties regardless of the point of measurement. This type of material is advantageous when high impact obliquities are anticipated.

Illuminating projectile – Used at night to illuminate the target, these projectiles contain an illuminant whose descent is haltered by a parachute.

Immunity zone – The period of impenetrability of the armoured citadel in terms of range (distance) against a specified attack. The inner limit of the immunity zone is secured by vertical armour. As range opens up, terminal energy of incoming projectiles decreases while their angle of fall increases. Consequently – if the vertical protection of the ship is strong enough to counter the specified threat – side armour becomes impenetrable beyond a certain range. The outer limit of the immunity zone is secured by horizontal armour. Albeit terminal energy decreases, impact angles against horizontal armour become more favorable for penetration as range opens up. Beyond a certain range – which marks the outer limit of the period of immunity – horizontal armour becomes vulnerable. The distance between the inner and outer limits is the zone where the ship is theoretically secured against the specified threat. The immunity zone constantly oscillates as the ship is pitching, rolling, its speed and relative bearing changes etc., which is why a too small period of immunity – in terms of distance – is inadequate to provide reliable protection. Ideally, armoured ships seek to engage at ranges corresponding to the center of their immunity zone. When the immunity zone of an armoured ship was calculated, the specified threat was often the gun carried by the ship itself, as the performance of foreign guns was not known in detail.

Impact angle – Same as angle of attack.

Initial velocity (IV) – Velocity of the projectile at the point of leaving the muzzle.

Kinetic energy upon impact – Kinetic energy of the projectile at the point of impact.

Limit velocity (VL) – Same as ballistic limit velocity.

Low-pressure gun – Refers to guns with below-average average pressure profile. Guns having an average pressure profile of less than about 9.0 tons/in^2 are termed low-pressure guns in this study.

Low-velocity gun – Refers to guns with below-average muzzle velocity. Guns having an initial velocity of no more than about 2,500 fps (762 mps) are termed low-velocity guns in this study.

Main battery – Guns of the largest caliber carried by the ship.

Maximum ordinate – Highest point of the trajectory.

Muzzle energy – Kinetic energy of the projectile at the point of leaving the muzzle. $E_k = \int p(x)dx * d^2 * \pi/4 = p_{av} * L * d^2 * \pi/4 = m*v^2/2$,

whence

E_k=kinetic energy,
d=diameter of gun barrel,
m=mass of projectile,
v=velocity of projectile,
p=pressure,
p_{av}=average pressure,
L=Length of gun barrel.

Muzzle velocity – Same as initial velocity.

New gun – An unused gun with no bore enlargement propelling shells at designed muzzle velocity.

Non-cemented armour (NC) – Same as homogeneous armour.

Obliquity – Same as angle of attack.

Ogive – Forward curved section of a projectile limited by its shoulder.

Period of immunity – Same as immunity zone.

Pressure profile – Pressure as a function of distance or time inside the gun barrel.

Range – The distance from a station on one's own ship to the target or some other point.

Rate of fire – Rounds fired as a function of time.

Relative effectiveness (R.E.) factor – Dimensionless explosive constant indicating the potency of explosives as compared to TNT, which has a R.E. factor of 1.00. Higher R.E. indicates greater destructive potential. The relation between R.E. and potency is directly proportional, i.e. R.E. factor indicates the relative mass of TNT to which an explosive is equivalent.

Relative target bearing – The bearing of the target from the firing ship measured in the horizontal from the bow of one's own ship clockwise from 0 deg. to 360 deg.

Retardation – Velocity loss during the flight of the projectile due to the drag force of the medium it travels through. Total retardation in terms of velocity loss equals muzzle velocity minus terminal velocity.

Ricocheting – The tendency of projectiles to bounce off the target beyond a certain angle of impact, which varies with each individual attack/plate combination.

Salvo – Two or more shots fired simultaneously.

Secondary armament – In the case of capital ships, all guns except for those of the largest caliber.

Sectional density – Projectile weight divided by the square of bourrelet diameter (m/D^2).

Semi-armour-piercing projectile (SAP) – Designed to defeat moderately armoured targets, projectiles of this type have thicker metallic walls and carry less explosive as compared to HE type projectiles, but have thinner walls and carry more explosives as compared to AP type projectiles.

Sheath hardening – Projectiles having this type of hardness distribution have greater hardness on the surface than at the center. The surface hardness of sheath hardened projectiles drops markedly only abreast their cavity. The Americans and the Germans preferred this hardness distribution.

Shock sensitivity – Tendency of an explosive to detonate when subjected to shock.

Starshell – Same as illuminating projectile.

Striking velocity – Velocity of the projectile at the point of impact.

Target angle – The relative bearing of one's own ship from the target, measured in the horizontal plane from bow of the target clockwise from 0 deg. to 360 deg.

Terminal energy – Same as kinetic energy upon impact.

Terminal velocity – Same as striking velocity.

Time of flight – Elapsed time between the projectile leaving the muzzle and hitting the target.

Trajectory – The path that an object with mass in motion follows through space as a function of time.

Vertex – Same as maximum ordinate.

List of Tables

Table No.	Description	Page
Table 1.1	Armament Comparison	15
Table 1.2	Broadside Comparison	15
Table 1.3	Comparison of Ballistic Characteristics	16
Table 1.4	Armour Penetration Capacity Comparison Based on Bu. Ord. Sk. 78841	16
Table 1.5	Secondary Armament Comparison	24
Table 1.6	Secondary Armament Broadside Comparison	24
Table 2.1	EC 3/ter 'Gufo' Radar	25
Table 2.2	DEM	27
Table 2.3	Type 284P Radar	27
Table 2.4	Fire-Control Equipment Comparison	29
Table 3.1	Littorio – Armour Protection	42-43
Table 3.2	Richelieu – Armour Protection	44-45
Table 3.3	Littorio – Dimensions	46
Table 3.4	Richelieu – Dimensions	46
Table 3.5	Immunity of Littorio VS. Richelieu (French)	47
Table 3.6	Immunity of Littorio VS. Richelieu (U.S.)	47
Table 3.7	Immunity of Richelieu VS. Littorio	47
Table AA.1	General Characteristics	75
Table AB.1	Decapped Projectiles	78
Table AE.1	Richelieu's Secondary Armament Fire-Control System	85
Table AK.1	Major Caliber Hits Scored Against French Capital Ships	111
Table AL.1	Armour Penetration Capacity of Littorio I.	113
Table AL.2	Armour Penetration Capacity of Littorio II.	113
Table AL.3	Gunnery exercise results (1938-1941) – Littorio	113

List of Figures

Figure No.	Description	Page
Figure 1.1	Main Gun Comparison	17
Figure 1.2	Broadside Comparison	18
Figure 1.3	Comparison of Ballistic Characteristics	19
Figure 1.4	Littorio's Armour Penetration Capacity	20
Figure 1.5	Richelieu's Armour Penetration Capacity	22
Figure 1.6	Armour Penetration Capacity Comparison	23
Figure 2.1	Littorio's Main Battery Fire-Control Sensors	30-31
Figure 2.2	Richelieu's Main Battery Fire-Control Sensors	32-34
Figure 3.1	Longitudinal Sections	48-49
Figure 3.2	Cross Sections	50-59
Figure 3.3	Immunity Zones at 90 deg.	60
Figure 3.4	Littorio's Immunity Vs. Richelieu (French) (I.)	61
Figure 3.5	Littorio's Immunity Vs. Richelieu (U.S.) (I.)	62
Figure 3.6	Richelieu's Immunity Vs. Littorio (I.)	63
Figure 3.7	Littorio's Immunity Vs. Richelieu (French) (II.)	64
Figure 3.8	Littorio's Immunity Vs. Richelieu (U.S.) (II.)	65
Figure 3.9	Richelieu's Immunity Vs. Littorio (II.)	66
Figure 3.10	Immunity Comparison (I.)	67
Figure 3.11	Immunity Comparison (II.)	68
Figure 3.12	Immunity Graph (I.)	69
Figure 3.13	Immunity Graph (II.)	70
Figure AE.1	Jean Bart's Guns and Directors (I.)	87-89
Figure AE.2	Jean Bart's Guns and Directors (II.)	90-92
Figure AF.1	SF Radar Equipment	93
Figure AF.2	SA Radar Equipment	94-96
Figure AF.3	SG Radar Equipment	97
Figure AG.1	AP Projectiles	99-100
Figure AH.1	Jean Bart's Protection Scheme	101-105
Figure AI.1	Littorio's Main Turret Mount	107
Figure AI.2	Richelieu's Main Turret Mount	108
Figure AJ.1	Longitudinal Section (Littorio)	109
Figure AJ.2	Longitudinal Section (Richelieu)	110

List of Pictures

Picture No.	Description	Page
Cover	Vittorio Veneto at the battle of Cape Spartivento.	---
Picture 1.1	The battleship Roma displays her armament.	12
Picture 1.2	The French battleship Richelieu, photographed on 18 May 1944.	14
Picture 2.1	The main battery director of the Italian battleship Roma is turned to port. (1940)	26
Picture 2.2	The French battleship Richelieu, after refit in the U.S., in September 1943. Note that the foretop secondary armament director had been removed.	28
Picture 3.1	The Italian battleship Vittorio Veneto, photographed in 1940 before completion. Notice the external side armour of the battleship's citadel.	37
Picture 3.2	Richelieu en route to New York in 1943. Note the smooth hull of the battleship, revealing no sign of her internal side armour. This picture also shows that the inclination of the turret face plates from the vertical was no more than 20 deg., not 30 deg. as often cited	40

Bibliography

<u>Books</u>
Dulin Jr., Robert O. and Garzke Jr., William H., et al. *Battleships: Allied Battleships in World War II.* Naval Institute Press; 1st edition (November 1, 1980) 978-0870211003

Dulin Jr., Robert O. and Garzke Jr., William H., et al. *Battleships: Axis and Neutral Battleships in World War II.* Naval Institute Press; 1st edition (November 27, 1985) 978-0870211010

Campbell, N.J.M. *Naval Weapons of World War II.* Naval Institute Press; 1st edition (January 1, 1985) 978-0870214592

Jordan, John and Dumas, Robert. *French Battleships, 1922–1956.* Seaforth Publishing (September 17, 2009) 978-1591144168

Bagnasco, Ermingo and de Toro, Augusto. *The Littorio Class: Italy's Last and Largest Battleships 1937-1948.* Seaforth Publishing (June 1, 2011) 978-1848321052

<u>Admiralty Minutes</u>
ADM 281/31 – Admiralty Armour Investigation Program (1946-1950), Investigation No. 4, The Ballistic Performance of Spaced Armour Assemblies and the Effect of Varying the Arrangement of Plate Thickness

ADM 281/37 – Admiralty Armour Investigation Program (1946-1950), Investigation No. 10, The Ballistic Performance of Cemented and Face Hardened Deck Armour (200 lbs/sq. ft. and 240 lbs/sq. ft.) Under Attack by Decapped A.P.C. Shell

<u>Naval Proving Ground (Dahlgren, Virginia)</u>
NPG 5-47 – Ballistic Tests and Metallurgical Examination of Japanese Heavy Armor Plate
NPG 3-47 – Metallurgical Examination of Standard U.S. Armor-Piercing Projectiles

<u>Marine Nationale</u>
Batiment de ligne "Richelieu"
Batiment de ligne "Jean Bart"

<u>Online</u>
http://www.navweaps.com

Other Books in this Series

On 4 November 1937 the super battleship Yamato was secretly laid down at Kure Naval Arsenal. Less than three years later, on 27 June 1940, invoking the "escalator clause", the Americans laid down the first unit of their greatest battleship class, USS Iowa, at the New York Naval Yard. Lacking accurate intel, both navies were convinced that their newest battleship was second to none, and thus capable of overwhelming any foe in a gunnery duel. In the possession of the actual technical characteristics of the two ships, relying on primary sources and empirical data, we now take on, once again, one of the most hotly debated questions among naval analysts and enthusiasts ever since: Who was right?

On 24 May 1941 the German battleship Bismarck engaged in a gunnery duel with the flagship of the Royal Navy, the battlecruiser HMS Hood, and the newly commissioned King George V class battleship, HMS Prince of Wales. Within minutes, Hood found a watery grave beneath the devastating shellfire of the legendary German ship. Not much later, Prince of Wales made smoke and left the field of battle. But would it have been indeed so foolhardy to fight Bismarck alone?

Available on Amazon.com in Kindle eBook and paperback formats.

Also by this Author

This comprehensive reference book contains detailed tabulations and line drawings of the protection schemes and immunity graphs of Allied battleships and battlecruisers designed after the Washington Naval Treaty, including the following capital ship classes:

North Carolina, South Dakota, Iowa, Montana, Alaska, Nelson, King George V, Lion, Vanguard, Dunkerque, Richelieu, Project 21, Type A, Project 23 (Sovetsky Soyuz), Project 23bis, Project 24, Project 69 (Kronshtadt), Project 82 (Stalingrad), Design 1047;

and fragmentary information on UP 41, Project 23UN, Project 25, Project 64.

This comprehensive reference book contains detailed tabulations and line drawings of the protection schemes and immunity graphs of Axis battleships and battlecruisers designed after the Washington Naval Treaty, including the following capital ship classes:

Littorio, Scharnhorst, Bismarck, H-39, O, Yamato;

and fragmentary information on, H-40, H-41, H-42, H-43, H-44 and B-65.

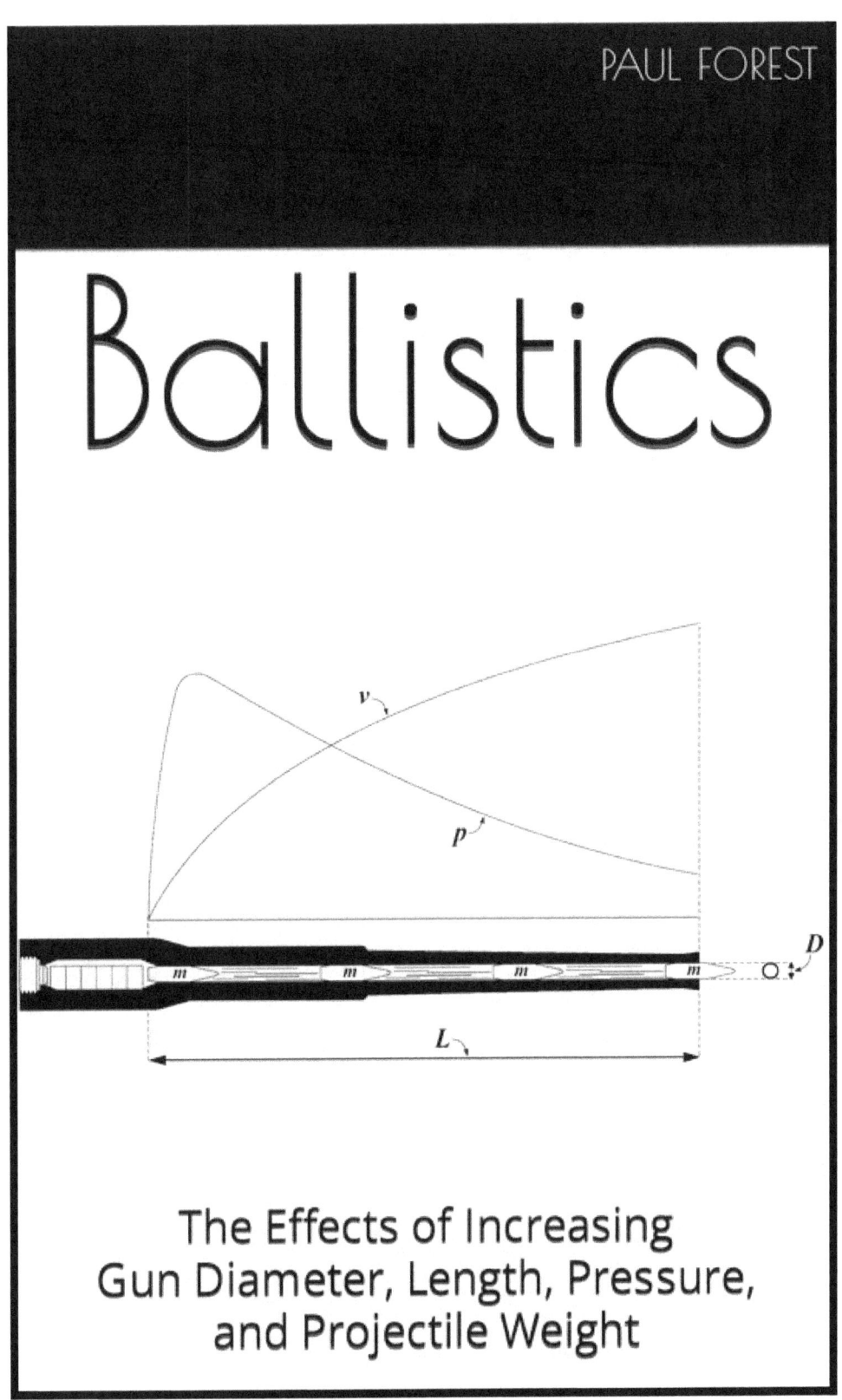

This short essay investigates the effects of increasing the diameter, length and pressure of large caliber naval guns, and the weight of projectiles, on ballistic characteristics and armour penetration capacity.

www.ingramcontent.com/pod-product-compliance
Lightning Source LLC
Chambersburg PA
CBHW080547220526
45466CB00010B/3065